Karl Shuker

The Menagerie of Marvels

A Third Compendium of Extraordinary Animals

Typeset by Jonathan Downes,
Cover and Layout by The Captain for CFZ Communications
Using Microsoft Word 2000, Microsoft Publisher 2000, Adobe Photoshop CS.

First published in Great Britain by CFZ Press

The Centre for Fortean Zoology,
Myrtle Cottage,
Woolfardisworthy,
Bideford, North Devon
EX39 5QR

ISBN: 978-1-909488-20-5

DEDICATION

To my dear mother, Mary D. Shuker (1921-2013), whose lifelong interest in wildlife guided and encouraged my own from my earliest days. Thank you for filling my world with wonder, joy, and love for such a long and very happy time. How I miss you, and how I wish that you were still here with me today and always. God bless you, little Mom.

The mother's heart is the child's schoolroom
Henry Ward Beecher – *Life Thoughts*

LIST OF CONTENTS

ACKNOWLEDGEMENTS

I wish to offer my sincere thanks to the following persons and organisations for their greatly-valued assistance, interest, and contributions in relation to my preparation of this book.

Silvana Pellegrini Adam; David Alderton; Raymond Bell; Markus Bühler; Igor Burtsev; Prof. John L. Cloudsley-Thompson; Martin Cotterill; Ben Coult; Jonathan Downes/CFZ/CFZ Press; Miroslav Fišmeister; John Gaughan; Markus Hemmler; Shaun Histed-Todd; Dr David Kirschner; Sir Christopher Lever; Oll Lewis; Juliet Lilienthal; Cameron A. McCormick; Raphaël Marlière; Carl Marshall; Tim Morris; Richard Muirhead; Dr Darren Naish; Nigel; Mark North; Hodari Nundu; Andy Paciorek; Michael Playfair; Lee Raiter; William Rebsamen; Roger C. Reeves; Robert Schneck; the late Mary D. Shuker; *Shropshire Star*; Susan R. Stebbing; Lars Thomas; Dr Warren D. Thomas; Robert Twombley; the late Gerald L. Wood.

My especial gratitude goes to Anthony Wallis, a longstanding friend and superb artist, for very kindly preparing this book's absolutely spectacular wraparound cover illustration, gorgeously showcasing a very formidable pair of terror birds - thank you so much, Ant!

I have sought permission for the use of all illustrations and substantial quotes known by me to be still in copyright. All illustrations not accompanied by a credit are assumed to be in the public domain. Any omissions or errors brought to my attention will be rectified in future editions of this book.

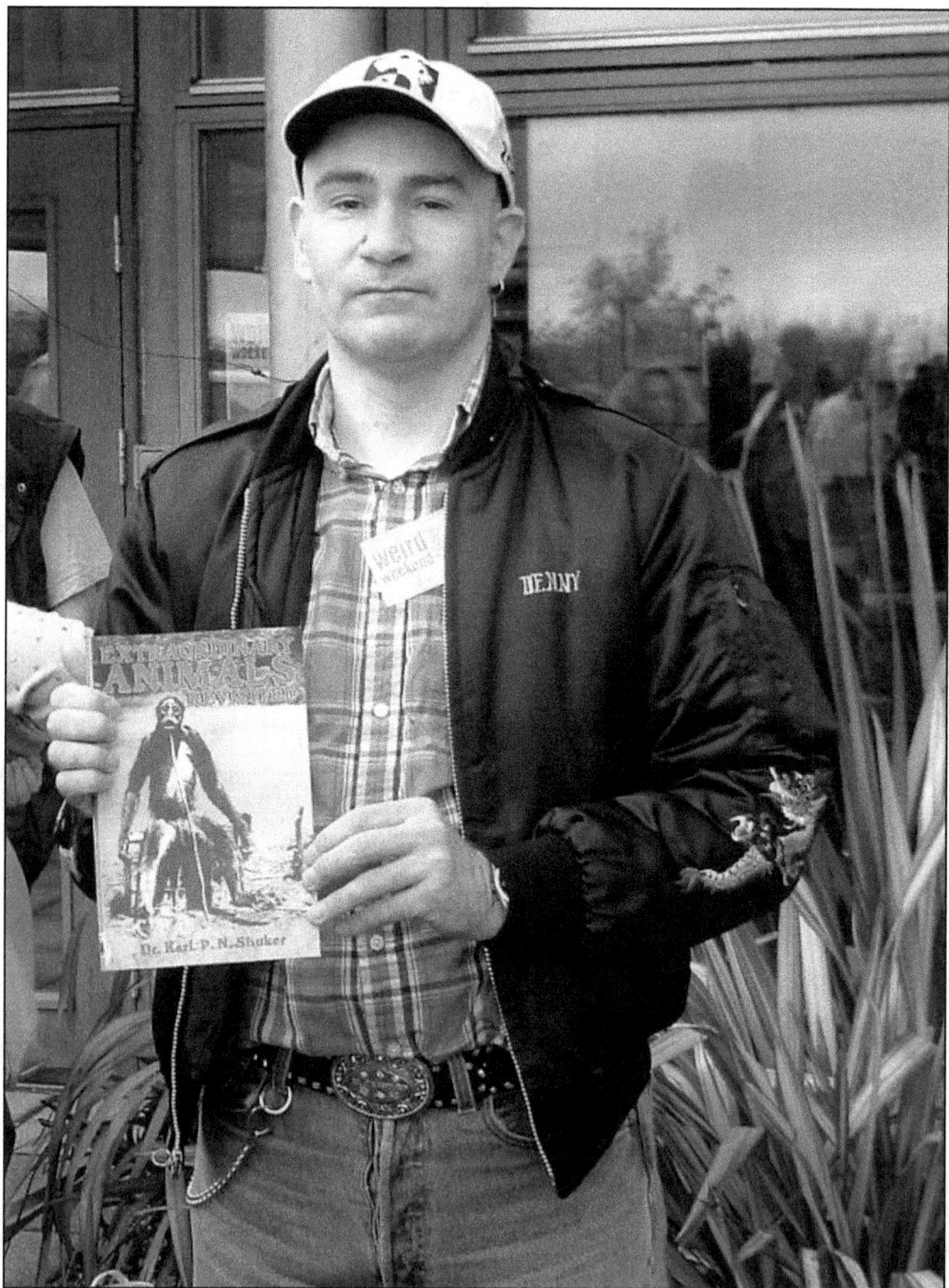

The author holding *Extraordinary Animals Revisited* during its official launch at the CFZ's 'Weird Weekend 2007' conference staged in August 2007 (Mark North)

INTRODUCTION

In all things of nature there is something of the marvellous.

Aristotle
– *De Partibus Animalium*, Book I, 645.a16

It was back in 1991 when my second book, *Extraordinary Animals Worldwide*, was published. Containing accounts of lesser-known cryptozoological beasts and scarcely-known mainstream creatures, and plentifully supplied with exquisite antiquarian chromolithographs, engravings, and other vintage illustrations wherever possible, it purposefully recalled a bygone generation of natural history books, dating predominantly from the 19th and early 20th Centuries, whose subject matter, generally a deft, eclectic interweaving of speculative zoology, the history of animal discovery, and wildlife mythology of the ancients, was popularly referred to by its authors and readers alike as romantic zoology.

Such was the enduring appeal of my book's modern-day contribution to this now all-but-lost subject – indeed, eventually gaining a cult status among cryptozoological aficionados in particular – that I was encouraged to prepare a much-expanded, updated edition, entitled *Extraordinary Animals Revisited*, which was published by CFZ Press in 2007. Updating some of the most popular chapters from the original book and also adding many new ones, it went on to attract an even greater following than its predecessor, and remains in print today.

By 'mixing and matching' cryptozoology and animal mythology with mainstream zoology, these two books have each enabled me to include within a single volume a much greater diversity of creatures than in other works of mine, and in turn have indulged me in my desire to investigate and write about certain truly obscure animals that have long fascinated me. Consequently, it was only a matter of time before I would give into temptation and compile a third compendium of extraordinary animals - and here it is, published once again by CFZ Press.

Making a welcome return within its pages is a much-updated, greatly-expanded version of the very popular chapter on terror birds (phorusrhacids) that originally appeared in *Extraordinary Animals Worldwide*. Ditto for the chapter devoted to those three overtly odd birds that we know as the shoebill, hammerhead, and boatbill. Also reappearing here is the giant rat of

Sumatra, which uniquely features in all three volumes (but in much greater detail here than in either of its predecessors). Otherwise, however, this latest book's chapters are not drawn from either of the two previous compendia. Instead, they are based upon a wide range of published articles of mine and various online posts from my ShukerNature blog, suitably expanded and updated whenever appropriate - but, faithfully continuing the extraordinary animals tradition, they are amply supplied with beautiful period illustrations and other eyecatching pictures throughout. They are also supplemented by a detailed bibliography and an index of animals documented in this book.

So, welcome one and all to my Menagerie of Marvels! Where else would you encounter fairy armadillos and go-away birds, roc feathers and a werewolf paw, a park of monsters in Italy and mystery beasts in the Vatican, whale-headed pseudo-pterodactyls and hammer-headed lightning birds, beech martens in Britain and winged toads in France, an invisible catfish and a dicephalous kestrel, reverse mermaids and the music of Ogopogo, earth hounds, moonrats, kinkimavos, lavellans, nandinias and Nandi bears, ajolotes, bristle-heads, mammoths, hoopoes, gorilla-sized man-eating baboons and giant rhino-eating terror birds, lake monsters and sea serpents, leopons and lizard cryptids, vermiform rock-slicing laser gazers, and so much more too, all within the scenic yet comfortingly-secure confines of a single book?

Enjoy your visit, and return whenever you wish – my menagerie's unique collection of extraordinary zoological esoterica and inexplicabilia will always be here to mystify and mesmerise you anew. You have only to step inside...if you dare!

Chapter 1:

FROM BIS-COBRA TO KUMI
- IN PURSUIT OF LESSER-KNOWN
MYSTERY LIZARDS

Snakes of all kinds are held in great horror by the natives of India, and they slay indiscriminately and ruthlessly all they come across, but this horror pales before the terror inspired even by the names of the bis-cobra and goh-sámp,— terror so great, that, if met with, the harmless animals are given the widest berth possible, and their destruction is never attempted. Though actual animals, they are virtually mythical, that is as regards the deadly properties assigned to them, and we easily recognise in them the originals of the flame-breathing dragon and deadly basilisk. The gaze of the bis-cobra is awful even from a distance and its bite is instant death; and if the goh-sámp breathes upon, or at you, you fall dead at once.

H.F. Hutchinson – *Nature*, 9 October 1879

Certain mystery lizards or lizard-like creatures have received a fair amount of publicity over the years, both within and beyond cryptozoological circles. These include a wide range of alleged giant monitors from around the world, various Congolese enigmas of the reptilian variety, and a controversial chameleon known as the Oldeani Monster (all of which I have documented in previous books). However, there are also a number of much less familiar examples on file, yet which are every bit as curious and compelling as their more famous counterparts. Consequently, this present chapter presents a diverse selection of these lesser-known mystery lizards.

THE BIS-COBRA – INDIA'S VENOMOUS LIZARD THAT NEVER WAS
In villages across the length and breadth of India even today, there remains tangible fear concerning a creature that may be small in size but is gargantuan in terms of the terror that the mere sight of it generates. Known most commonly as the bis-cobra, according to generations of fervently-believed native folklore and superstition this modest-sized Asian lizard has such a

venomous bite that anyone so inflicted will die instantly. Needless to say, no species matching this description is known to science. Yet there is no doubt that the bis-cobra does exist. So what precisely *is* this noxious entity, and how can these contradictions be resolved?

The name 'bis-cobra' (or 'biscopra'), which is used most prevalently in western India, loosely translates as 'venomous cobra'. Bearing in mind that all cobras are venomous, this is a particularly direct, hard-hitting way of emphasising just how exceptionally toxic this animal is – or, to be more accurate, allegedly is.

I first read about the bis-cobra many years ago, when perusing a delightful, humorous book on Indian wildlife entitled *The Tribes On My Frontier*, published in 1904 and written by 'EHA' - the pen-name of Indian amateur naturalist and artist Edward Hamilton Aitken (1851-1909). His description of it summarises very succinctly the basic attributes of this mysterious reptile:

> But of all the things in this earth that bite or sting, the palm belongs to the biscobra, a creature whose very name seems to indicate that it is twice as bad as the cobra. Though known by the terror of its name to natives and Europeans alike, it has never been described in the proceedings of any learned society, nor has it yet received a scientific name. In fact, it occupies much the same place in science as the sea-serpent, and accurate information regarding it is still a desideratum. The awful deadliness of its bite admits of no question, being supported by countless authentic instances; our own old *ghorawalla* [horse-keeper] was killed by one. The points on which evidence is required are – first, whether there is any such animal as the biscobra; second, whether, if it does exist, it is a snake with legs or a lizard without them. By inquiry among natives I have learned a few remarkable facts about it, as, for instance, that it has eight legs, and is a hybrid between a cobra and that gigantic lizard commonly miscalled an iguana [in India, 'iguana' is a term popularly misapplied to monitor lizards]; but last year a brood of them suddenly appeared in Dustypore, and I saw several. The first was killed by some of the bravest of my own men with stones, for it can spring four feet, and no one may approach it without hazard of life. Even when dead it is exceedingly dangerous, but, with my usual hardihood, I examined it. It was nine inches long, and in appearance like a pretty brownish lizard spotted with yellow. It has no trace of poison-fangs, but I was assured that an animal so deadly could dispense with these. If it simply spits at a man his fate is sealed.

After some effort, EHA finally managed to capture a bis-cobra alive in his own garden using a butterfly-net, much to the great consternation of his native butler, watching the proceedings from a considerable distance. He then kept it for a time as a pet, without suffering any adverse effects.

Pre-dating EHA's account by several decades, however, was a lengthy report on the bis-cobra by a Mr John Grant that featured in the inaugural volume of the *Calcutta Journal of Natural*

EHA's own bis-cobra drawing in *The Tribes On My Frontier*

History, published in 1840. In it, Grant referred to a specimen of a reputed bis-cobra specially captured for him to examine. Approximately 6 in long, it was attractively patterned with irregular streaks of small bead-like markings of alternating dark and light grey colour. Anxious to observe its lethal effect, Grant introduced a mouse into the glass container housing this lizard. But far from the mouse meeting a rapid demise, it fought spiritedly with the lizard for a short time, each biting the other, before the two combatants retreated to opposite sides of the container, neither of them appearing any worse for their savage encounter. So much for the bis-cobra's virulent venom.

In her book *East of Suez*, Alice Perrin included an eventful incident in which a European living in India demonstrated dramatically but beyond any doubt that the bite of a bis-cobra was harmless. He achieved this by lifting a brown and yellow specimen out of a pot in which it had been trapped, and then, when it seized hold of his hand with its teeth, holding it up, still biting him, for all of his horror-stricken native helpers to witness. After anxiously waiting for a while to watch their doomed master's fully-anticipated demise, they finally dispersed when it became clear that he was totally unharmed. Perrin also documented an encounter with a tree-climbing bis-cobra, measuring about 14 in long.

EHA was not the only source of native lore claiming that the bis-cobra doesn't even have to bite in order for its venom to prove lethal. Several other writers have also alleged that it only has to spit a single drop onto someone's skin for its potency to prove instantaneously fatal, searing through the victim's skin, entering their bloodstream, and eliciting certain death. Indeed, in his book *Indian Peepshow* (1937), Henry Newman was even assured by a local man that if this dire lizard so much as spat at a tree (let alone a person), a hole would burn right through and the tree would die.

In parts of India outside the western zone where reports of the bis-cobra are most rife, this supposedly deadly beast is conflated with another mysterious but equally malign reptile called the hun khun. Likened to a small slow-moving lizard with a fat tail, much the same powers of venomous potency are attributed to it as to the bis-cobra, but even the blood of the hun khun is reportedly toxic, and its skin reputedly contains lethal poison glands.

As science knows of no species corresponding to the bis-cobra, how can this enigmatic lizard be explained? Down through the decades, several different zoological identities have been proposed for it. In their book *Venomous Reptiles* (1969), Sherman and Madge Minton proposed that the bis-cobra was the East Indian leopard gecko *Eublepharis hardwickii*, a small

Fat-tailed or common leopard gecko *Eublepharis macularius*

stout species with a noticeably thick tail. Of course, as this gecko (like all others) is wholly harmless, in order to accommodate its identification as the bis-cobra the latter's dread reputation as a highly venomous creature must necessarily be nothing more than native superstition and folklore. The Mintons identified the hun khun as the closely related fat-tailed or common leopard gecko *E. macularius*.

Henry Newman noted that in the hotter, drier parts of India, 'bis-cobra' was a term applied to a rarely-seen, fleet-footed species of grey lizard. And that in Eastern Bengal, it is an elusive crested lizard occasionally spied in gardens and on walls.

In his newly-updated two-volume encyclopaedia of cryptozoology, *Mysterious Creatures* (2013-14), George Eberhart noted the Mintons' view. He also speculated that an alternative explanation for the bis-cobra is that it is a non-existent composite beast, created by locals combining (and sometimes confusing) reports of venomous snakes with non-venomous lizards.

By far the most commonly-held view as to the bis-cobra's identity today, however, is one that, ironically, had originally been suggested by a number of authors more than a century ago. In his earlier-mentioned book, for instance, EHA recalled how his captured bis-cobra gradually grew larger until within a few weeks it had developed into an unmistakeable 'iguana' (i.e. monitor lizard). He concluded dryly:

> Some people would jump to the conclusion that it was a young iguana
> to begin with. My butler would endure the thumbscrew sooner.

Similarly, in his above-documented *Calcutta Journal of Natural History* report from 1840, John Grant concluded that the specimen which he had pitted against the mouse was nothing more than a young goshamp – a local name in present-day Bangladesh and West Bengal for the common Indian (Bengal) monitor *Varanus bengalensis*. This species is widely distributed throughout the Indian subcontinent, and whereas adults are mainly terrestrial, juveniles are more arboreal, thereby explaining reports of tree-climbing bis-cobras. Their spotted patterning also matches morphological descriptions of the bis-cobra.

Moreover, writing in *Beast and Man in India* (1891), John Lockwood Kipling stated:

> The large lizard, *varanus* [sic] *dracaena*, which is perfectly innocuous,
> like all Indian lizards, is called the bis-cobra by some.

Varanus dracaena is a synonym of *Varanus bengalensis*. Kipling actually owned a pet specimen of this monitor species, and whenever he held it he was invariably warned of its supposed deadliness by native observers.

Monitors were also confidently identified as the bis-cobra by L.S.S. O'Malley in his book *Bengal, Bihar, Orissa and Sikkim* (1917). In *Eye in the Jungle* (2006), acclaimed Tamil writer M. Krishnan affirmed that the dreaded bis-cobra has been shown by naturalists to be nothing more than young, harmless specimens of the common Indian monitor – a statement confirmed in the standard work on this varanid species, Walter Auffenberg's monograph *The Bengal Monitor* (1994). Today, therefore, the term 'bis-cobra' is treated merely as a synonym for the latter monitor.

The real bis-cobra - a juvenile common Indian monitor *Varanus bengalensis* (Jayendra Chiplunkar/Wikipedia)

All that remains to explain now are supposed cases (such as that of EHA's ghorawalla) in which a bite from a bis-cobra, i.e. a young Indian monitor lizard, caused a person's death. If such cases are indeed genuine, how are such deaths possible, bearing in mind that *V. bengalensis* is not venomous?

Various possibilities come to mind. For instance, in recent years it has been shown that certain varanids, notably the Komodo dragon *V. komodoensis*, do actually possess venom glands. Although the venom produced by them is not normally fatal to humans, someone exceptionally sensitive to it may conceivably suffer anaphylaxis in a manner comparable to the response of certain people to the venom in bee or wasp stings. And even if no venom is present, the bacteria present on the teeth of these lizards could readily infect a wound created by a bite from one, and thus induce septicaemia. Moreover, the superstitious fear generated by the bis-cobra may in itself be sufficient to bring about death by heart failure in someone bitten by a monitor lizard.

It may well be that isolated incidents involving one or more of these causes of death following a bite from a technically harmless lizard were sufficient to engender the tenacious myth of the lethal bis-cobra, especially among medically-untrained villagers. It would also explain the diversity of lizard species claimed by them to be the bis-cobra at one time or another. The ultimate result was a fascinatingly Frankensteinian creation of the deadliest lizard that never actually existed.

THE RIFT VALLEY'S MYSTIFYING MUHURU

One summer's day during the early 1930s, missionary Rev. Cal Bombay and his wife Mary were driving through the Muhoroni region of Kenya's Rift Valley, journeying towards Nairobi, when they observed a remarkable object blocking their path ahead. They had to brake sharply to avoid hitting it, and saw that it was an extremely large lizard-like beast, lying across

Muhuru, based upon eyewitness descriptions (William Rebsamen)

the road, apparently sunbathing. It was at least 10 ft long and dark grey in colour, with four stubby legs, and a snake-like head. Most eyecatching of all, however, was the series of large diamond-shaped serrations running from the rear portion of its head along the entire length of its back, becoming smaller in size as they neared the tip of its tail, and greatly resembling the jagged spines or scales often depicted on the back and tail of dragons in medieval art.

Both aghast and alarmed at what they were seeing, the Bombays wisely decided to stay inside their car until after about 20 minutes the creature rose up and slowly plodded away into the brush, its body staying close to the ground. Enquiries revealed that this mystifying reptile was well known to local hunters, who referred to it as the muhuru. Were it not for its dorsal diamond-shaped spines, the muhuru could conceivably be reconciled with an exceptionally large monitor lizard, but no known species of varanid possesses these anomalous serrations.

A THREE-TOED ODDITY FROM EAST AFRICA
The muhuru is reminiscent of another still-unidentified African lizard, which was once sighted by a big game hunter in the border region between Ethiopia and Sudan, and briefly documented in 1952. According to the hunter's report, this lizard was approximately 10-12 ft long, and dirty grey in colour, with a lizard-like head, but very loose skin, and a ridge of barbs or spines running along its back and tail. It moved via a waddling walk, sometimes sliding on its belly, but most distinctive of all was that each of this mystery reptile's four feet only possessed three toes. Each toe bore a large claw, and it left behind distinctive three-clawed footprints.

Australia is home to the yellow-bellied three-toed skink *Saiphos equalis*, but otherwise three-toed lizards tend to result from individual embryonic maldevelopment or from postnatal injuries in which one or more toes have been lost. In any case, this mystery lizard's ridge of dorsal and caudal spines coupled with its odd mode of locomotion are strange enough characteristics in themselves to make impossible any attempt to identify it with known species.

ADDITIONAL AJOLOTES?
Most closely related to teiids, microteiids, and lacertids, worm-lizards are also known as amphisbaenians. They are predominantly fossorial, spending much of their time underground (though they do appear on the surface sometimes during heavy rain), and the vast majority are limbless, outwardly resembling large earthworms or small snakes.

However, there are four species of very unusual, unmistakeable worm-lizards called ajolotes (a name sometimes confusingly and incorrectly applied to axolotls too). The best known species is the common ajolote *Bipes biporus* (the other three are *B. alvarezi*, *B. canaliculatus*, and *B. tridactylus*). All are pink or greyish-purple in colour, and all are instantly distinguishable from every other amphisbaenian by virtue of their small but distinctly-formed pair of forelimbs. Positioned just behind the ajolotes' head, these limbs are equipped with a pair of conspicuously broad, clawed feet that superficially look like large ears!

19th-Century engraving of an ajolote

Currently, ajolotes are known only from northern Mexico's Baja California peninsula. However, there are reports of odd creatures greatly resembling these two-legged worm-lizards both to the north and to the south of their official distribution range - as revealed here, in what is the most comprehensive account ever published on this fascinating subject.

In a *Copeia* paper published on 10 December 1938, eminent Kansas University herpetologist Dr Edward Harrison Taylor revealed that during the summers of 1928-1930 and again in 1934, he had conducted searches in southeastern Arizona's Huachuca and Santa Catalina mountain ranges for a mysterious reptile that he believed to be a species of ajolote. Although he failed to discover it, he did collect a number of intriguing eyewitness accounts that all described an undeniably ajolote-like cryptid.

The account that had first inspired Dr Taylor to conduct his searches was that of a placer-gold miner, who informed him that he had occasionally dug out of the sand and gravel along the small creek in the Huachuca Mountains' Ash Canyon "a small snake 10-14 inches long with two small legs near its head. They were purple or brown in color". Needless to say, this is an excellent verbal portrait of an ajolote.

Taylor visited an elderly naturalist called Dr Biedermann in a canyon close to Ash Canyon; Biedermann had lived in the Huachuca Mountains for over 30 years. He informed Taylor that he believed these mountains were home to "a rare *chirotes* [a name derived from the now-obsolete ajolote genus *Chirotes*, having been superseded by *Bipes*]", and he also claimed to own a preserved specimen of it. Unfortunately, no such specimen could be found when he and Taylor inspected his personal reptile collection.

While collecting reptile specimens himself on Mount Lemon in the Santa Catalina mountain range, Taylor stopped at a small hotel there, whose owner, a Mrs Westbrook, told him about a very unusual pet that she had kept for three months after finding it one evening in the hotel's garden during a rainstorm. She referred to it as a snake but stated: "It had a pair of legs coming out where its ears should be". Sadly, this remarkable beast had escaped, never to be seen again, but several people who had observed it when it was captive vouched for the hotel owner's story. Its description clearly recalls an ajolote.

In 1934, while on Mount Lemon, Taylor had met a Mr Doty of the Forest Service, who informed him that several months earlier some of his workers had killed three two-legged snakes while removing piles of rocks in order to drill post holes for telephone poles. Doty took Taylor to the exact spot, and they spent a day there searching for more of these creatures, but none was discovered. However, it was an extremely dry day, whereas it had been raining on the day that Doty's workers had found their specimens.

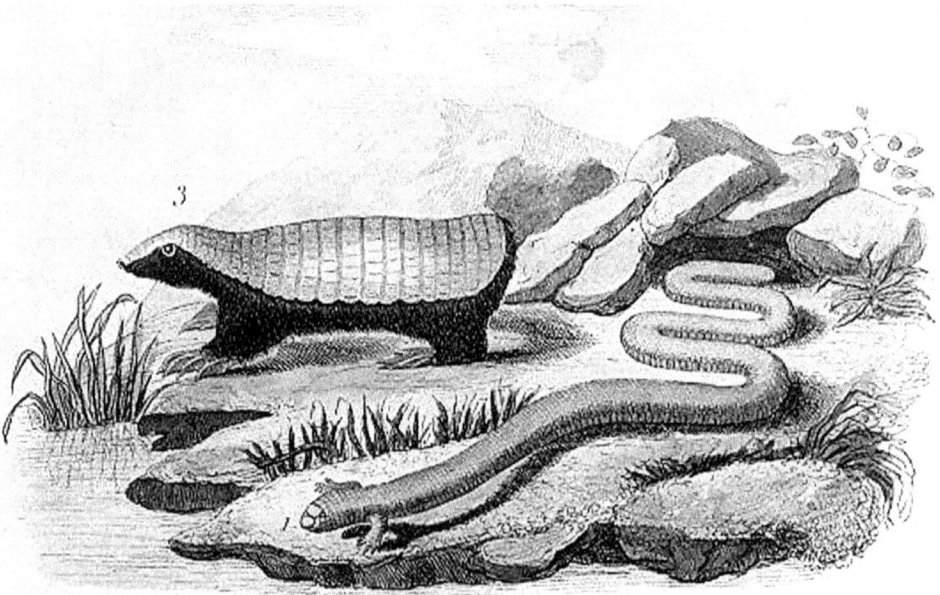

Ajolote and pichiciego (see Chapter 10), from *Dictionnaire Pittoresque d'Histoire Naturelle* by Félix Édouard Guérin-Méneville (1836)

Moreover, the forest guard at Mount Lemon's outlook station told Taylor that he had found in the Huachucas what he described as a snake with "two legs on its neck. It was lavender and white below. The legs were so short that it didn't use them to walk on". Regrettably, although he collected snake skins as a hobby, the guard hadn't preserved this specimen's, because it was so short.

So precise and consistent are all of the above reports that it certainly seems very plausible that a form of ajolote, conceivably representing a fifth, presently-undescribed species, does indeed still await discovery in Arizona's Huachuca and Santa Catalina ranges.

Equally notable is the following description published in 1823 of alleged ajolotes encountered on 27 June 1820 in the South Platte Valley somewhere between Ogallala in Nebraska and Julesburg in Colorado (i.e. possibly near today's Nebraska-Colorado border), during an expedition from Pittsburgh to the Rocky Mountains in 1819-1820 by order of the Hon. J.C. Calhoun, the USA's Secretary of War, under the command of Major Stephen H. Long. The description was penned by Edwin James, the expedition's geologist and botanist, and can be found within Vol. 1 of the two-volume *Account of an Expedition from Pittsburgh to the Rocky Mountains*, compiled by James from the notes of Major Long, the expedition's zoologist Thomas Say, and various other expedition members:

> We observed in repeated instances, several individuals of a singular genus of reptiles (Chirotes. Cuv.) which, in form, resemble short serpents, but are more closely allied to the lizards, by being furnished with two feet. They were so active that it was not without some difficulty that we succeeded in obtaining a specimen. Of this (as was our uniform custom, when any apparently new animal was presented) we immediately drew out a description. But as the specimen was unfortunately lost, and the description formed part of the Zoological notes and observations, which were carried off by our deserters, we are reduced to the necessity of merely indicating the probability of the existence of the *Chirotes lumbricoides* [a now-obsolete ajolote binomial] of naturalists, within the territory of the United States.

The zoological notes were in saddlebags stolen by three deserters from the expedition's small military escort towards the end of the expedition. Sadly, they were never recovered, despite strenuous efforts to do so.

Some herpetological researchers have subsequently disparaged this account, variously claiming that the location specified in it would be too cold in winter to harbour ajolotes, suggesting that the creatures' behaviour seemed too lively for them to be ajolotes, or questioning the accuracy of the creatures' identification as ajolotes. Other researchers, conversely, have treated it in a more positive manner, and of particular interest is that some modern-day reports of possible ajolotes have emerged from Nebraska and Colorado, as now revealed here.

While searching for lizards on the ranch of A.J. Simants of Tryon, Nebraska, during a herpetological collecting trip spanning 11-12 June 1951, biologist Dr Howard A. Dundee from Louisiana's Tulane University was informed by Mr Simants and his 13-year-old son that this area was home to a lizard with only two legs, positioned close to its head. It allegedly crawled like a snake, was greyish or silver-greyish in colour, and measured 1 ft or so in length. Found along fence posts or fence rows, it was most commonly encountered by the Simantses when they were excavating for fence posts or pulling up rotting fence posts. Dr Dundee was taken to some refuse piles where such lizards had been seen, but none made an appearance.

Ajolote (bottom right) and various lizards in an engraving from 1894

In addition, Dr T. Paul Maslin's annotated checklist of Colorado's herpetological fauna, published in 1959 within *University of Colorado Studies, Series D (Physical and Biological Sciences)*, contains two latter-day eyewitness reports of possible ajolotes. One was obtained by Dr Maslin in 1946 from a well-educated janitor at Colorado State University (CSU) in Fort Collins, who described in minute detail the animal that he had discovered some years earlier on a farm owned by him, located southeast of Fort Collins, and which unquestionably recalled a *Bipes* specimen. When Dr Maslin showed him a specimen of *Eumeces multivirgatus*, a short-limbed quadrupedal skink but superficially ajolote-like and native to that area, the janitor rejected it at once, stating that this skink was common and well known, whereas what he had seen was very different. The other eyewitness report, obtained by Maslin shortly after the janitor's, was from a graduate CSU student, whose description of a creature found under a rock on the west side of a hogback a few miles west of Loveland was another convincing account of an ajolote.

In or around 1955, some 35 form letters that included a sketch of *Bipes* were circulated to biology teachers, wildlife officers, and others in northeastern Colorado where such creatures may have been seen, but they elicited no positive response. Could it be that there were indeed ajolotes in Nebraska and Colorado in earlier times but that these have now died out? Or do they still exist there but, just like their known Mexican counterparts, are resolutely reclusive?

Even within Mexico itself, there is at least one report of a putative ajolote occurring well beyond these legged amphisbaenians' confirmed distribution limits. On 25 October 2012, I received an email from a Mexican correspondent who informed me that a few years earlier, a government biologist specialising in medically-significant venomous animals and living in one of Mexico's southern states had told him that he had once found a small but very strange creature in his garden. The biologist stated that it was very similar in appearance to *B. biporus* both in form and colour, with mole-like forelegs clearly adapted for digging.

If ajolotes do indeed occur outside their officially recognised distribution range, and their existence is confirmed one day, this would be a very significant herpetological discovery. Yet in view of these legged worm-lizards' notoriously reclusive lifestyle, it would not be implausible either. Let us hope, therefore, that some herpetological enthusiast(s) will now take up the challenge posed by the tantalising reports documented here, and instigate without delay a thorough search for the elusive 'eared snakes' described in them.

THE CENAPRUGWIRION – WALES'S 'DAFT' FLY-CATCHER

Britain's rather meagre array of indigenous limbed lizards are all lacertids, but if the cenaprugwirion is more than Welsh whimsy this mystifying lizard-like creature may conceivably be something much more special.

The name 'flycatcher' normally conjures up images of the feathered variety, such as Britain's pied flycatcher *Ficedula hypoleuca* and spotted flycatcher *Muscicapa striata*. However, for a fair number of years the inhabitants in and around Abersoch in Gwynedd, North Wales, have been applying this same name to a very different – and much more mysterious - scaly species.

It even has its own local Welsh names – cenaprugwirion and genaprugwirion. The second of these names loosely translates as 'daft flycatcher' - 'gwirion' ('daft'), 'pryf' ('fly'/'insect'), and 'genau' ('mouth'). But if we assume that it isn't simply a local yarn, what exactly *is* Wales's 'daft' flycatcher? That remains the question.

It is generally likened by eyewitnesses to a large lizard, approximately 1 ft long, and muddy-brown in colour, with a head as big as an orange, very mobile eyes that apparently look as if they are forever rolling about in their sockets, a large tongue that it uses for catching flies, and a very noticeable dewlap (flap of skin beneath its chin). It lives in burrows in the ground or in earthen banks, where it spends much of its time, and was once said to be quite common here but since the 1980s has rarely been seen.

Needless to say, this enigmatic creature's morphology does not match that of any known species of reptile native to Britain. Consequently, unless it is a native species that has totally eluded formal scientific detection until now – which is surely unlikely on an island whose wildlife has been minutely studied for centuries – the cenaprugwirion must constitute a foreign species that has somehow established itself here, presumably after originally escaping (and/or having been released) from captivity, and long enough ago to have gained for itself some local names. We already play host, after all, to a very eclectic array of non-native exotica, from parakeets and wallabies to American bullfrogs and Aesculapian snakes – so why not a saurian interloper?

The fundamental problem is deciding the cenaprugwirion's taxonomic identity. It shares characteristics with a range of very different species, which explains why, when consulting with a number of herpetological experts, I have received a diversity of suggestions – including iguana, agama, skink, monitor lizard, and chameleon (this last-mentioned suggestion was no doubt influenced by the cenaprugwirion's very mobile eyes).

Unfortunately, however, the likelihood that any of these tropical lizards could even survive for very long in Great Britain's cool climate, let alone overwinter and eventually establish the kind of viable population formerly claimed for the cenaprugwirion by its human neighbours, seems very low. Furthermore, although all of the above-nominated lizard candidates display certain morphological or behavioural similarities to this Welsh mystery beast, none is comprehensively convincing.

However, one additional identity has also been suggested, and which has intrigued me very much, because it provides the closest fit to the cenaprugwirion in terms of both morphology and lifestyle. Yet it isn't actually a lizard at all, but is something much more startling.

With respect to unique fauna, New Zealand is famous for its multitude of endemic birds – and for the tuatara *Sphenodon punctatus* (sometimes split into two species). Although superficially lizard-like (and actually classed as an agama when first brought to Western scientific attention in 1831), the tuatara is in reality the only surviving member of an otherwise exclusively prehistoric reptilian lineage that constitutes an entirely separate taxonomic group from all others – the sphenodontians.

Engraving of tuataras, from 1901

Dinosaur contemporaries, the sphenodontians died out elsewhere millions of years ago, but the line leading to today's tuatara persisted to the present day in New Zealand because this dual-island country has no native mammals or large lizards that could have competed with and very conceivably wiped it out. Tragically, however, once Westerners reached New Zealand and began introducing Western predatory mammals such as cats, dogs, and rats, the tuatara's numbers rapidly fell. Today it exists principally on a series of small offshore islets and in captivity, having long been extirpated on the North and South Islands, although recent years have seen some success in reintroducing it onto North Island.

Is it even remotely conceivable, however, that long before this, the tuatara had already established itself as a reptilian alien far removed from its native New Zealand, i.e. within the environs of Abersoch, North Wales? Let's look at the evidence for such a bold claim.

Morphologically, the tuatara possesses the large head, long tongue, dewlap, and body size described for the cenaprugwirion. Furthermore, each of the tuatara's eyes is encircled by a very prominent bony ridge, making the eye sockets seem bigger than they really are, thereby emphasising the movements of its eyes. This in turn could explain why the cenaprugwirion's eyes appear to be moving continuously within their sockets. Also of note is that the tuatara occurs in a range of colours, from blackish-brown to olive green, and including the mud-brown shade named for the cenaprugwirion.

Behaviourally, the tuatara definitely exhibits the cenaprugwirion's burrow-dwelling preference and insectivorous diet. And the tuatara is renowned for being able to withstand low temperatures. Due to its very low metabolic rate, it is active in New Zealand at temperatures down to at least as far as 5°C (41 F), and hibernates during the winter, so it should have no problems with surviving Great Britain's climate. Equally significant is the tuatara's longevity - because this archaic reptile is a veritable Methuselah, and not just in palaeontological terms. Wild specimens take 20-30 years just to attain sexual maturity, they continue growing until they are 35 years old, and can live for over a century.

This extremely impressive lifespan is very important with regard to the tuatara's candidature as the cenaprugwirion, because it can shed some much-needed light on this extraordinary case's greatest paradox - how an exclusive (and nowadays exceedingly rare) reptilian New Zealander could possibly be thriving within the countryside of North Wales. The tuatara was not uncommon in the British pet and exotic animal trade during the Victorian era, so although it is not likely, it is not impossible either that a few specimens (perhaps no more than a single pair) could have escaped or have even been purposefully released from captivity somewhere in North Wales. This was, after all, a notorious period in time for misguided attempts by many private individuals to introduce all manner of alien animals into the wild in Britain.

Yet, both physiologically and behaviourally, tuataras would be well-equipped to survive in their new surroundings - and not for just a few years. Many decades could pass, during which time they could attain maturity, breed, and thus establish a colony, in precisely the manner that they have done on their offshore isles in New Zealand. And as tuataras are primarily nocturnal, they would not attract undue attention from their human neighbours, except on sunny days when they would emerge from their burrows to sunbathe nearby, just as they do once again back home in New Zealand.

Needless to say, the concept of a naturalised tuatara colony living in North Wales in modern times must surely rate as one of the least likely cryptozoological cases on file. Then again, this is just what scientists said about sphenodontian survival...until the tuatara was discovered.

Consequently, if anyone reading this chapter lives in or near Abersoch, and has any sightings or recent information concerning the cenaprugwirion, I would love to hear from you, just in case it may still not be too late to solve the longstanding yet previously little-known mystery of North Wales's daft but (hopefully) not-departed saurian flycatcher.

A DRAGON LIZARD FROM ZIMBABWE

The 31 currently-recognised species of agamid belonging to the genus *Draco* are commonly referred to as gliding lizards or even as flying dragons, due to their paired, wing-like gliding membranes, formed from elongated ribs interlinked by thin folds of connecting skin. As far as mainstream science is concerned, they are all exclusively Asian in zoogeographical distribution. In February 1995, however, English cryptozoological researcher Richard Muirhead received a fascinating account from Sithembile Ncube, a pen-friend living in Bulawayo, Zimbabwe, concerning a seemingly unknown species of native Zimbabwean lizard

19th-Century engraving of *Draco* flying dragons

that sounds remarkably similar to a *Draco* flying dragon.

According to an article (source unknown to me) that had been read by Sithembile's brother, this mysterious lizard is brownish-grey in colour, with glossy black eyes, rough skin, and a not overly long tail that it will raise up vertically into the shape of a letter 'C' when sensing danger. Much more remarkable, however, is that it reputedly sports a pair of short wing-like structures composed of light bones and thin flesh-like material. When they are extended and the lizard flies (i.e. glides?), its legs are folded inward, i.e. pressed close against its undersides, and its

***Draco*-like lizard (at bottom right) on Ludolf's map of Abyssinia**

tail is rolled upwards. This extraordinary lizard is rarely seen as it is extremely shy, is active only during early morning and late sunset, and inhabits rocky mountainous areas. It feeds upon insects and smaller lizards.

It would be easy to dismiss this story as fiction, but the very fact that it was apparently derived from a published article as opposed to verbal hearsay makes it intriguing to me, and I would dearly like to read that article for myself. Also, there is a second, entirely independent, but astonishing piece of corroborative evidence for the presence of a winged lizard of this description in Africa.

In 1683, eminent German scholar and Ethiopian specialist Hiob Ludolf (1624-1704) included in one of his works a map of Abyssinia (now Ethiopia, but in Ludolf's time a term applied to a much greater expanse of Africa) that depicts a rhinoceros, a pair of unmistakeable African elephants – and what looks *exactly* like a *Draco* lizard with its familiar gliding membranes

28

fully extended!

Is this a case of Ludolf having confused his continents with respect to *Draco*? Or is it a tantalising iconographical clue that such a reptile really is indigenous to Africa?

And as a most exciting postscript: after I'd informed him in September 2013 about Zimbabwe's flying mystery dragon-lizard, English pet expert/author David Alderton passed the details on to an African herpetologist friend, Paul Donovan. And in October 2013, Paul replied with the remarkable news that not only is the Zimbabwean wife of one of his friends familiar with this creature, but she even knows the area (within a nature reserve) in which to find them! Could there be a major zoological discovery about to be made in Zimbabwe? I'm currently awaiting further news from Paul via David, so let us all hope that it is positive.

THE NGARARA AND KUMI OF NEW ZEALAND
Although more than 60 species of lizard (all geckos and skinks) are native to New Zealand, as is the outwardly lizard-like tuatara (see this chapter's cenaprugwirion section), none even attains 4 ft in length. In contrast, Maori traditions and also sightings by European settlers here testify to the onetime and possibly even modern-day existence of substantially larger lizard forms.

One of these is the ngarara, which was generally claimed to be larger than the tuatara (which is up to 28 in long), boasting a girth of 8-16 in. It also sported a serrated dorsal crest, as well as prominent 3-4-in-long teeth that caused the upper lip to project forward slightly. It frequented manuka scrub-covered areas of South Island's Canterbury Plains, where it was said to inhabit burrows. Could this have been an extra-large form of tuatara?

Bigger still was the kumi, which was claimed to be at least semi-arboreal. In September 1898, a Maori bushman working at a station near Gisborne on North Island was allegedly confronted by a 5-ft-long specimen of this mystery lizard that advanced upon him before losing its nerve and fleeing into a rata tree. A team of fellow workers as well as the station's owner set out in search of the kumi, but although they didn't locate it they did spy some footprints that they duly photographed.

A drawing of a kumi was even given to Captain James Cook when he arrived at Queen Charlotte Sound in 1773, by a local Maori chief called Tawaihura. He informed Cook that these giant lizards lived in trees and were greatly feared by his people. Having said that, certain subsequent native reports given to early European settlers claimed that the Maoris hunted and ate such lizards. Perhaps this explains why their species (possibly an arboreal varanid?) has never been identified or even confirmed – for could it be that the kumi was hunted into extinction before science had even received the opportunity to recognise its existence?

A MONSTER LIZARD FROM BRAZIL
Equally startling was the monstrous lizard-like creature witnessed at close range by Swedish

explorer Harald Westin one day in 1931. While travelling down Brazil's Marmore River, he observed an extremely odd-looking reptilian beast pacing along the nearby shore. About 20 ft long, it possessed an alligator-like head, and four lizard-like feet, but its body reminded him in overall shape and form of a distended boa constrictor. As his boat approached it, the creature raised its head to look at him, revealing a pair of small, scarlet-coloured eyes, and Westin became so frightened that he shot it with his gun. However, the bullet did not seem to have any injurious effect upon this bizarre animal, because it simply ambled away in a totally unconcerned manner.

The creature's distended body may indicate that it had recently eaten a large meal, but there are insufficient additional details to enable its likely zoological identity to be ascertained. Moreover, as an experienced explorer like Westin would certainly have been familiar with this region's known crocodilians, such as caimans, it seems unlikely that anything as prosaic as those latter reptiles could explain the Marmore monster.

If nothing else, this eclectic selection of reports indicates that the herpetological world's catalogue of species may still contain a number of sizeable gaps needing to be filled. Or, at the very least, that there are some little-known yet very curious accounts buried deep within the cryptozoological archives that would greatly benefit from some modern-day scientific re-examination, and also if possible some belated pursuits in the field, just in case...

Chapter 2:
KINKIMAVO AND BRISTLE-HEAD
– A COUPLE OF MAINSTREAM MYSTERY BIRDS

Also classified tentatively with the orioles is the little-known Kinkimavo (*Tylas*) of the Madagascan forests.

Placed tentatively with the starlings as a third subfamily, the Pityriasinae, is the rare and little-studied Bristle-head of Southern Borneo forests. Although it had long been classified with the helmet shrikes (p. 270) because of its peculiar head feathering, and sometimes placed in a family by itself, what little is known of the Bristle-head's behaviour and habits has led most students today to regard it as a highly aberrant starling.

Oliver L. Austin Jr and Arthur Singer – *Birds of the World*

J ust because a species has been formally named and described doesn't mean to say that it is no longer mysterious. Both of the birds documented in this chapter were scientifically recognised during the 1800s, yet remain as ornithologically controversial and generally obscure today as they were back then – but they have fascinated and tantalised me ever since childhood.

TWO LIFE-CHANGING, LIFE-AFFIRMING BIRD BOOKS

I owe my abiding love and knowledge of birdlife to two truly wonderful books that were bought for me by my family when I was a child, and which I still own today. Both were very big and both were exquisitely illustrated throughout in full colour. The first of these life-changing volumes, which I received when I was around 5 years old, was *The Colourful World of Birds*, written by Jean Dorst, filled with lavish paintings by Pierre Probst, and published in 1963 by Paul Hamlyn of London. Its many chapters were themed around habitat, behaviour, nesting and breeding, migration, and interactions with humans. Although aimed primarily at older children, its contents were detailed and highly informative, and introduced me to such avian marvels as the quetzal, dodo, megapodes, birds of paradise, hummingbirds, and which kinds of birds were to be found in which types of habitat.

Holding the two bird books that have influenced me most (Dr Karl Shuker)

When I was about 8 years old, I received the second epochal bird book, which was a truly magnificent, sumptuously-illustrated tome entitled *Birds of the World*. First published in 1961 and once again by Paul Hamlyn of London, it was written by Oliver L. Austin Jr, and was packed with countless spectacular paintings by Arthur Singer. Even today, it remains one of the most beautiful bird books ever published, as well as a classic, milestone work within the ornithological literature – and for me, this enormous book, which seemed almost as big as I was on that fateful day when I first laid eyes upon it in Beatties department store in

Wolverhampton, West Midlands, it was love at first sight! When my mother took pity on my forlorn face after we discovered that this wondrous publication was priced at what was in those days a veritable king's ransom for a book – 5 guineas!! (£5.25 in decimal currency) – and bought it for me anyway, I was rendered speechless with delight, and hugged it closely to me throughout our journey back home on the bus.

It is no exaggeration to say that *Birds of the World* transformed and expanded my knowledge concerning the taxonomy and diversity of birds to a degree not even remotely approached by any other publication that I have ever read since. For whereas the contents of *The Colourful World of Birds* were divided into the various thematic categories noted above, *Birds of the World* was a comprehensive taxonomic survey of our world's avifauna, presenting each taxonomic order in turn and within it each family, accompanied by gorgeous illustrations of representative species from every one, with over 700 species illustrated in total. Moreover, whereas *The Colourful World of Birds* only included common names, *Birds of the World* also presented the scientific binomial name for every species illustrated (as well as for many that were only referred to in the text).

Suddenly, my young brain was ablaze with images and facts concerning strange, exotic, and frequently multicoloured birds from every corner of the globe, often with strange names and even stranger life histories. Over countless re-readings of this magical book, I familiarised myself with the likes of peppershrikes and bellmagpies, tyrant flycatchers and false sunbirds, currawongs and curassows, todies and tropic-birds, ioras and o-os, tinamous and tapaculos, jacamers, frogmouths, puffbirds, vangas, spiderhunters, umbrellabirds, kagus, greenlets, drongos, phainopeplas, hemipodes, mesites, and much much more – including the kinkimavo and the bristle-head.

Their odd-sounding names alone would have been enough to incite my curiosity, but this curiosity was heightened by the facts that little seemed to be known about either of them, that both of them had long perplexed ornithologists concerning their taxonomic affinities, and that neither of them was illustrated nor even described morphologically in what had become my veritable bible of ornithology, *Birds of the World*.

Today, I could have readily sought information and illustrations concerning these birds online. Back in the 1960s and right up to the late 1990s, conversely, the pre-internet world in which I lived presented much greater difficulties in obtaining data concerning two such obscure species, especially during my childhood and teenage pre-university years. And so, for a long time I remained tantalised and tormented in equal measures by the sparse details provided in *Birds of the World* relative to the kinkimavo and the bristle-head.

Indeed, its account of the kinkimavo was no more than a single line at the bottom of p. 222, ending the section devoted to Oriolidae, the passerine family housing the Old World orioles and figbirds. I have quoted that lone line at the onset of this chapter in my own present book.

I have also quoted there the brief coverage on p. 275 of *Birds of the World* concerning the bristle-head, which constituted the final paragraph in that book's section on Sturnidae, the starling family.

During the decades that have passed since the publication of *Birds of the World*, avian taxonomy has experienced many revolutions, not least the dramatic changes postulated by genetic studies. These have revealed hitherto-concealed affinities between such externally-dissimilar taxa as the New World birds of prey or cathartids and the storks, and the reclassification as flycatchers of a number of familiar species traditionally deemed to be thrushes (such as the European robin, nightingale, redstarts, and chats). Inevitably, such studies have also continued to engender speculation and dissension concerning the true taxonomic affinities of the kinkimavo and the bristle-head.

Thanks to the internet and access to all manner of specialist works during and since my university days, I was eventually able to flesh out the bare bones of information provided by *Birds of the World* for these twin birds of mystery, as well as to track down images of them. At last, the kinkimavo and bristle-head have emerged from the shadows of ornithological obscurity, unveiled for me in all their quirky but no less compelling glory. So here is what they look like, and what I have learnt about them.

MADAGASCAR'S MYSTIFYING KINKIMAVO
Let's begin with the kinkimavo,

(Top) Kinkimavo - painting by Joseph Wolf, accompanying this species' official description in the *Proceedings of the Zoological Society of London*, 1862; (Bottom) Bornean bristle-head – painting from 1838

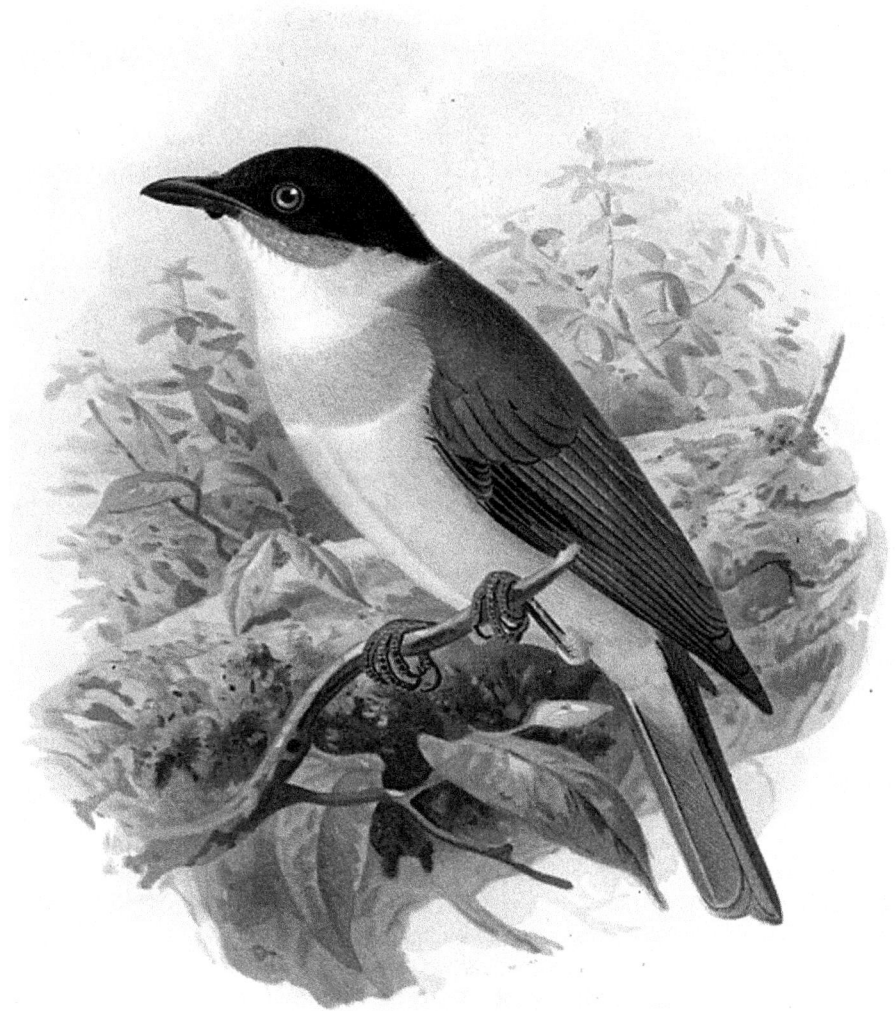

The kinkimavo - painting by J.G. Keulemans, 1880s, in *Histoire Physique, Naturelle et Politique de Madagascar*, by Alfred Grandidier and Alphonse Milne-Edwards

which was scientifically described on 13 May 1862 by German ornithologist Dr Gustav Hartlaub within the *Proceedings of the Zoological Society of London* (pp. 152-153). He formally dubbed it *Tylas eduardi*, in honour of its discoverer, Sir Edward Newton, who was, in Hartlaub's own words:

> ...a gentleman who has recently visited Madagascar, and whose zealous efforts have very materially forwarded our knowledge of the ornithology of the East-African archipelago.

(For trivia fans: during his time as a colonial administrator in Mauritius from 1859 to 1877, Sir Edward was also the person who sent to England what became the type specimen of the dodo!)

Its generic name, *Tylas*, was derived, somewhat oddly, from the Greek word 'Tulas', referring to a kind of thrush mentioned by ancient scholar Alexander Myndios, even though the kinkimavo bears scant (if any) resemblance to one. As for its unusual common name, this is one of several local names given to it by tribes sharing its Madagascan homeland.

Roughly 8 in long, and the only member of its genus, the kinkimavo is a sedentary, insectivorous species that exists as two readily-distinguished subspecies. The nominative *T. e. eduardi* is the much more common form, and occurs in primary rainforest and sometimes adjacent second growth too in eastern Madagascar. However, *T. e. albogularis*, named after its characteristic white throat (and deemed by some researchers to warrant reclassification as a separate species in its own right), is much rarer, found only in certain local areas of dry forest and mangroves in western Madagascar.

The kinkimavo is primarily very dark brown/black and white in colour. However, it also has an orange tinge to its underparts, plus a grey-green tinge to its upperparts, upper wings, and tail. There is no notable sexual dimorphism. The breeding season extends through autumn, and into as far as January in the nominative subspecies, after which a small cup-shaped nest of leaves and moss is constructed high in a tree, and two eggs are laid in it.

As for the kinkimavo's taxonomic position: originally, it had been placed by Hartlaub within the bulbul family, Pycnogonidae, but it was subsequently reassigned to the orioles, as noted in *Birds of the World*. However, the current consensus is that like a number of other mystifying passerine species endemic to Madagascar, the kinkimavo is actually a vanga. Apart from one species that has extended its range into the nearby Comoro Islands, the vangas are found only in Madagascar, and constitute a little-known taxonomic family, Vangidae, that contains an extremely diverse assortment of species (22 or thereabouts in total number, depending upon which researcher is consulted!). The more conservative members are outwardly shrike-like, and in earlier days the vangas were deemed to be shrikes and thus were referred to as vanga-shrikes (as in *Birds of the World*). Certain others resemble and behave like warblers or babblers. However, they also include some much more extreme species.

A pair of sickle-billed vangas, 1880s painting by J.G. Keulemans

Most notable among these are the sickle-billed vanga *Falculea palliata*, whose long curved beak is reminiscent of a wood hoopoe's; and the extraordinary helmet vanga *Euryceros prevostii*, which sports a huge casque-bearing arched beak. Much less distinctive externally but extremely deceptive is what was once known as the coral-billed nuthatch but is now called the nuthatch vanga *Hypositta corallirostris*. On account of its great outward similarity to the nuthatches, this small grey bird with the bright red beak was long deemed to be one itself, but later studies exposed it as a vanga in disguise. And now the kinkimavo appears to be yet another member of this surprising bird family, and is thus frequently referred to lately as the tylas vanga (although to my mind this is a much clumsier, less memorable name than the infinitely more euphonious kinkimavo).

The vangas constitute an excellent example of adaptive radiation. They appear to have evolved from a single ancestral form that, after reaching Madagascar several million years ago, rapidly diversified in form to occupy via segregated speciation a number of vacant ecological niches present there. So dramatic is the degree of morphological radiation exhibited by the vangas (and especially by their range of beak shapes, mirroring their wide range of feeding preferences), in fact, that if Charles Darwin had chosen to visit Madagascar rather than the

A painting from 1831 of the helmet vanga

Galapagos Islands, he would have encountered a far more elaborate example of evolutionary diversity with the vangas than the version exhibited by the Galapagos finches that inspired and shaped his Theory of Evolution.

THE BAFFLING BRISTLE-HEAD OF BORNEO

And now to the second member of this chapter's pair of mainstream mystery birds: the Bornean bristle-head. A very distinctive species, it was formally described even earlier than the kinkimavo, in 1835, by eminent Dutch zoologist Coenraad Jacob Temminck, who named it *Barita gymnocephala*. Four years later, it was assigned its own genus by French ornithologist René Primevère Lesson, and is now known as *Pityriasis gymnocephala*. Its generic name is actually a skin disease of the scalp, characterised by warts upon a bald head, and its specific name translates as 'bald-headed'. Both names refer to this bird's partly-naked, warty-skinned head and the unusual but characteristic bristles borne upon it.

Whereas the kinkimavo is demurely monochrome, the bristle-head is unabashedly gaudy. Roughly 10 inches long, it sports a massive hooked black beak, and its sombre black/dark grey wings and body plumage contrast markedly with the bright red hue of its head, neck, throat, and thighs, plus its white wing patches, and the very odd-looking skin projections or bristles, pale yellow in colour, borne upon its naked, warty crown. They earn this species its common name, and resemble bare feather shafts.

A taxiderm specimen of the Bornean bristle-head

Endemic to the island of Borneo, the bristle-head is an uncommon species, categorised as Near Threatened by the IUCN, and is sparsely distributed throughout its lowland forests (primary and secondary) and mangrove swamps. It feeds upon large invertebrates, small vertebrates, and fruit, and often associates within the forest canopy in mixed flocks with a range of other bird species (as does the kinkimavo in Madagascar). Breeding behaviour is largely unknown, and only a single oviduct egg has ever been recorded with certainty from this mysterious bird.

As with the kinkimavo, the bristle-head's taxonomic status has generated much contention ever since its scientific description, but unlike the former bird's it still does so today. Over the years, the bristle-head has been variously assigned to the helmet shrike family (Prionopidae), the woodswallow family (Artamidae), the crow family (Corvidae), the starling family (Sturnidae), and the Australian butcherbird and currawong family (Cracticidae). In 1951, Drs Ernst Mayr and Dean A. Amadon created a brand-new family especially for it, Pityriaseidae, in which most researchers still house it, albeit as much a placing of convenience than one of certainty. Most recently, however, there have been suggestions that this baffling bird should be rehoused in a new family, Tephrodornithidae, which contains the equally perplexing flycatcher-shrikes (genus *Hemipus*) and the woodshrikes (genus *Tephrodornis*).

Bearing in mind that the woodshrikes are also deemed to be closely allied to the vangas, this could actually mean that the bristle-head and the kinkimavo are themselves related – one final, highly-unexpected twist to the much-tangled taxonomy of these two little-known yet abidingly-fascinating avian enigmas.

Incidentally: as a teenager fired with enthusiasm for investigating all manner of cryptozoological and neo-cryptozoological subjects, I was rash enough one day to mention the kinkimavo to a group of friends. Its name caused much merriment and no shortage of ribald comments, not least of which was the enquiry from one friend as to who this kinky Mavo was, and where could he meet up with her? In view of this, I was very thankful that I'd had the good sense not to mention the bristle-head!!

Chapter 3:
THE BRITISH BEECH MARTEN - A MUSTELID MYSTERY

It is an indisputable fact that, whereas a hundred and fifty years ago there were two species of marten recognised in Britain, only one has ever made it into the history books, and it also seems reasonable that utilising cryptozoological methodology, giving credence to eyewitness reports, and to the etymological evidence, the people who were actually familiar with the creatures considered them to belong to two separate species, which seems to be valuable circumstantial evidence pointing towards them being two separate species.

Jonathan Downes – *The Smaller Mystery Carnivores of the Westcountry*

Two species of marten are common in continental Europe. One of these is the pine marten *Martes martes*, characterised by its lithe build, dark pelage, and yellow throat patch. The other is the beech or stone marten *M. foina*, characterised by its slightly smaller size but stockier build, lighter pelage, and white throat patch. However, only the pine marten is officially native to Britain. Yet as will now be revealed, until as recently as the late 19th Century many naturalists and natural history books claimed that the beech marten also existed here – so why is it no longer deemed to have done so?

As long ago as 1429, it was against the law in Scotland for anyone other than a lord or a knight to wear clothing made from or trimmed with the fur of mertriks (pine martens) or funyies (beech martens). Ironically, however, what may be the earliest published literary reference to the beech marten allegedly existing in Britain can be found not in any work of natural history but in a famous play by a certain William Shakespeare! In Act iv, scene 4 of *Romeo and Juliet* (1597), Lady Capulet refers to Capulet as a 'mouse-hunt' – an odd-sounding term that is actually one of several alternative English names formerly applied to the beech marten (others being martern, marten cat, marteron, and martlett).

It was Welsh naturalist Thomas Pennant (1726-1798), however, who first popularly claimed the beech marten for the U.K. within the zoological literature. In his book *British Zoology*

Illustration from 1833 depicting the beech marten (top) and pine marten (bottom)

(1768), he readily distinguished between the pine marten and the beech marten, but listed both as inhabitants of Britain, though he alleged that the latter species was much less common in England than the former. Pennant reiterated his categorisation of the beech marten as a British species in his subsequent tome *History of Quadrupeds* (1793).

In his own book, *A Natural History of British and Foreign Quadrupeds* (1841), James Hamilton Fennell also claimed both the beech and the pine marten for the U.K. However, he held the converse view to Pennant as to these species' relative abundance, stating that the pine marten was rarer than the beech marten in Great Britain as a whole, but that in England neither was found except in the northern parts. As for Scotland, he stated:

> The beech marten does sometimes [occur], in the highlands of Scotland, where it is common and called tuggin...however, it is now nearly exterminated in the south of that country.

Fennell additionally stated that in Selkirkshire, the beech marten has been seen descending to the shore at night to feed upon molluscs, in particular the large basket mussel. Clearly, therefore, he had no doubt whatsoever that this species of marten was indeed a fully-recognised member of the British fauna.

As noted above, Fennell claimed that in England the beech marten occurred only in its northern portion, but not everyone apparently agreed with this. In 1877, for instance, the *Transactions of the Devonshire Association* published a paper on the mammals of Devon by Edward Parfitt in which not only the pine marten but also the beech marten was included, the latter species under the name of 'marten cat'. Its entry was brief but informative:

> This species is now, I believe, nearly extinct as a systematic war is waged against it by preserves of game. Mr. P.F. Amery informs me that the last he has heard of was killed near Ashburton about six years ago [i.e. c.1870].

Nor is this the only mention of the beech marten formerly existing in southwestern England. In his fascinating little book *The Smaller Mystery Carnivores of the Westcountry* (2006), to which I wrote the foreword, Devon-based CFZ founder Jonathan Downes included several other 19th-Century reports, referring to this species' apparent presence in Somerset, Dorset, and Cornwall too.

A particularly interesting example was the report that occurred in a 1903 paper by C.W Dale on Dorset mammals published in the *Proceedings of the Dorset Natural History and Antiquarian Field Club*. For whereas the pine marten received only a single line of text coverage in this paper, the beech marten (referred to once again as the marten cat) boasted an entire paragraph - indicating not merely that it did exist in this county but also that it was actually the better-known marten species there:

> The Rev. William Chafin, in his "Anecdotes of Cranborne Chase," records marten-cats as one of the sort of animals hunted there, but believes them

(1816) to be nearly extinct, their skins being too valuable for them to be allowed to exist. In 1836 one was caught alive near Stock House by the Rev. H.F. Yeatman's hounds, but, biting the huntsman's hands severely, it was kept alive for some little time.

In view of such a quantity of testimony seemingly substantiating the reality of the beech marten's presence within the U.K., why today, therefore, has it been entirely expunged from the British fauna, not only as a present-day member but even as an erstwhile one? The answer is a paper from 1879 authored by British zoologist Edward R. Alston, and published in the *Proceedings of the Zoological Society of London.*

Illustration from 1886 depicting the pine marten (top) and the beech marten (bottom)

Entitled 'On the Specific Identity of the British Martens', Alston's paper confirmed that the pine marten and beech marten were indeed separate species (in the past, some authorities had suggested that they were merely varieties of a single species), and he provided a very detailed, diagnostic set of distinguishing cranial, dental, and external characteristics for each of them. However, after then providing a summary of some previous published coverages referring to the two species existing in Britain, Alston queried whether there was any firm evidence for the presence here of the beech marten.

In particular, he stated uncompromisingly that during the previous ten years, he had taken every opportunity to study every British marten specimen that he could, from England, Wales, Scotland, and Ireland, in the precise hope of uncovering any that were beech martens, but not a single one had proved to belong to that species – they had all been pine martens. He also noted that fellow English zoologist Edward Blyth had conducted similar investigations, with the same result, and added that he had never encountered any knowledgeable wildlife observer acquainted with the diagnostic differences between beech and pine marten who had ever seen an authentic beech marten obtained in Britain.

Furthermore, Alston pointed out that the most visible external difference between the two species – the colour of their throat patch – could be very deceptive. This was because in old specimens of pine marten, the yellow hue had often faded to white, thereby matching that of the beech marten - a potential source of confusion among non-specialists, who may well have mistakenly assumed that they had observed bona fide beech martens when in reality they had simply seen aged pine martens. Faded museum specimens could also have been responsible for promoting this scenario of mistaken taxonomic identity.

Alston's paper was thus the death knell for the British beech marten. Additionally, in subsequent years palaeontological evidence has indicated that the beech marten's westerly spread across Europe from its earlier southern European, Middle Eastern, and Central Asian heartland may well have occurred too late for it to reach Britain before the English Channel formed, around 200,000 years BP (Before Present). This sizeable sea channel split Great Britain off from mainland Europe, turning it into an island and thus denying the beech marten direct overland access into Britain. Nevertheless, in his authoritative book *The History of British Mammals* (1999), noted mammalogist Dr Derek Yalden offered a crumb of hope for supporters of the British beech marten by noting that there are not many sub-fossil British-derived specimens of marten on record, with the possibility ever present of a new specimen yielding a surprise.

Also worth considering is Downes's telling statement in his own book, noting that Alston was denying all the historical records of an animal that had been well known to generations of naturalists, trappers, hunters, and churchwardens. Surely not every single one of these persons, some of whom were in constant contact with such animals, had been mistaken down through the centuries?

Ultimately, however, the only conclusive means of verifying the onetime presence of the beech marten in Britain would be to uncover a confirmed specimen – but perhaps we already have done! There are two very enigmatic taxiderm martens in existence that may conceivably shed some very illuminating light upon this longstanding zoo-historical mystery.

One of these is a stuffed, cased marten currently on display in a small museum within the Square and Compass pub in the Dorset village of Worth Matravers. What is intriguing about it is that it lacks the small bib portion of the throat patch that pine martens (but not beech martens) often exhibit. Of course, there is a great deal of individual intraspecific (as well as interspecific) variation in the shape, colour, and extent of the throat patch in martens. So this is not diagnostic, but it is definitely intriguing.

The stuffed mystery marten from the Square and Compass pub in Worth Matravers, Dorset (Mark North)

So too is the second specimen, which has only recently been brought to cryptozoological attention and has not previously been documented in any book. When not working at the Stratford Butterfly Farm, naturalist Carl Marshall, from Bidford-on-Avon in Warwickshire, is a longstanding cryptozoological enthusiast, and while wandering around a car boot sale held at Stratford-upon-Avon one Saturday in early July 2013 he saw an odd-looking taxiderm marten priced at just £10 on one of the stalls. Although its fur was very faded, it was otherwise in decent condition, but appeared stockier in build than Carl expected for a pine marten. So he asked the seller where it had originated, and was startled to learn that it had come from Dorset!

Being mindful of the alleged former existence of beech martens in Dorset and elsewhere in the Westcountry, Carl lost no time in purchasing this tantalising specimen, and as both his father and various family friends are professional taxidermists, Carl hopes to restore it to a good state of preservation. Moreover, as they agree that it does seem burlier than pine martens tend to be,

Carl Marshall and his stuffed mystery marten (Dr Karl Shuker)

Carl has carefully taken hair samples from several different areas of its body and has forwarded them to zoologist Lars Thomas at Copenhagen Zoological Museum for formal hair analysis in the hope of determining its taxonomic identity. And what if it should turn out to be a beech marten? Carl showed me this intriguing specimen when I visited him on 6 September 2013, and also kindly promised to provide me with the results of the hair analysis as soon as he receives them from Lars – so watch this space!

Finally: A third taxiderm specimen of a putative British beech marten is also on record, but unlike the previous two documented here, its current whereabouts are unknown. As uncovered back in 2009 by Welsh Fortean researcher Richard Holland, the August 1887 issue of a long-defunct Welsh journal entitled *Bye-gones* contained a fascinating account of the capture and subsequent preservation of an alleged beech marten in Palé, which is a large estate containing extensive deciduous woodlands, located in what is now the county of Conwy. The account was penned by a writer identifying themselves only as 'J. L1', and it reads as follows:

> A short time ago whilst one of Mr Robertson's [Palé] watchers was going round his beat he found to his surprise in one of the steel traps set for vermin, a ferocious looking animal, and it was with the greatest difficulty he could approach him. Being caught by the hindleg, he was springing

forward, and striving his utmost to bite him. The watcher, however, managed to set another trap, so his forefeet were secured, and he was taken alive to the keeper's residence.

It was sent to Mr Shaw of Shrewsbury to be preserved, and Mr Shaw described him thus: "He is a fine example of the Beech Marten (Martes foine [sic]) and has become very rare; they reside in trees, and feed mostly on squirrels. When descending to the ground they destroy rats and other small animals."

We may also add that in size he is smaller than a fox, and of a dark brown colour with white along the breast, the tail and head very much resembling the cat. He has been well stuffed and looks very well, placed on a branch of a tree. This can be seen at Mr Thomas Hughes's, Glyn, Llandrillo (the keeper's house).

Could the taxidermist have been mistaken regarding this specimen's taxonomic identity, or was a genuine beech marten caught and killed in Wales less than 130 years ago? Only its taxiderm avatar can answer that telling question, and perhaps one day it will – always assuming that it still survives, that it can be successfully tracked down, and that it can then be subjected to formal scientific scrutiny. How likely that this crucial sequence of events will ever take place, however, is another matter entirely.

Chapter 4:
GO-AWAY BIRDS
– WHY THEY DON'T, AND WHAT THEY DO

[The grey go-away bird] is a very common and most conspicuous bird in the Kruger National Park, its characteristic call note – a nasal 'go away! go away!' – keeping everybody aware of its presence. The game animals, having long ago realised that they have nothing to fear of man, pay very little attention to it...[but] in areas open to hunting, the animals react exactly as the deer of a European forest will react to the harsh call of the jay...The white-bellied go-away bird has a very wide range of bleating, barking, mewling, chuckling and cackling sounds, and some African tribes credit it with imitating the voices of all sorts of animals to deceive the hunters...[The bare-faced go-away bird's] call is similar to that of the grey go-away bird: 'go-ah, go-ah'; and, as in the other two species, a party is apt now and again to break into shrieks of laughter.

John Gooders (ed.) – *Birds of the World* (9 vols)

A bird cried from one of the trees.
'Listen,' said Mum. 'The loerie.'
We heard the loerie every day. It was a nondescript grey thing, with a haunting, high-pitched shout. 'Go 'way, go 'way!'
I had known ever since I could remember that the loerie was the Go-Away Bird. It only just occurred to me that its song could be any number of things – it depended on how you listened.
'Why is it called the Go-Away Bird?' I asked Mum.
'Because it says Go Away.'
'But...' I thought about it for a second, then remained silent. I thought it was significant that we had always heard it as 'Go Away', every day, all these years.

Amelia Eames – *The Cry of the Go-Away Bird*

G o-away birds – they sound like something from a children's fantasy novel, akin perhaps to the Never bird in J.M. Barrie's classic *Peter Pan*. And yet they are indisputably – and very audibly – real. Ask any native African hunter attempting to sneak up on his intended animal victim when one or more of these birds is perched close by.

They earn their onomatopoeic name from the sound of their extremely loud, oft-repeated cry, which does sound rather like "g-away!" – check out this YouTube video (http://www.youtube.com/watch?v=sBexB-c8_rI) of some grey go-away birds in fine voice, and listen for yourself.

Acting very much as self-appointed wildlife sentinels, go-away birds perch high up in tree tops, well out of danger themselves, and then proceed via their raucous alarm cries to warn any unsuspecting antelope or any other prey animal in the vicinity of approaching threats, such as human hunters, lions, cheetahs, and other predators. As a result, both prey and predator do indeed go away – the grateful prey to live another day, the seriously aggrieved predators to glare with impotent rage at their feathered nemeses.

But what *are* go-away birds, zoologically speaking? In fact, they are unusually plain-plumaged relatives of the typically gaudy touracos (and less gaudy plantain-eaters). Touracos, go-away birds, and plantain-eaters (all of which are often referred to colloquially as loeries) are famous for their controversial classification – sometimes allied with the cuckoos, sometimes accorded their very own taxonomic order, Musophagiformes. And they are equally celebrated for producing not one but two porphyrin-derived pigments found nowhere else in the animal kingdom – turacin and turacoverdin.

Grey go-away bird, by Claude Gibney Finch-Davies, 1916

Turacin is the bright crimson pigment commonly seen edging their wings (in all other birds, red colouration is due to the presence of carotenoids). And as its name indicates, turacoverdin is a green pigment that is liberally exhibited in touraco plumage (whereas in all other birds, green colouration is due to a combination of yellow carotenoids and the scattering of blue light caused by the prismatic qualities of their feathers' surface structure).

In stark contrast, go-away bird are garbed only in the most sombre, nondescript greys and whites, but possess showy crests and long tails, and average a total length of 20 in or so. Like typical touracos, they are endemic to tropical Africa, gregarious, non-migratory, and live principally upon fruit and flowers. However, they inhabit open or less forested country, whereas other touracos tend to be predominantly forest dwellers.

White-bellied go-away bird (Steve Garvie/Wikipedia)

Bare-faced go-away bird, by Joseph Smit,
1881

Three species are recognised. The most attractive member of this trio is the white-bellied go-away bird *Corythaixoides leucogaster*. Very widely distributed in East Africa, it boasts a lengthy tail handsomely barred in black and white, and bright white underparts too.

Less dramatic is the aptly-named grey go-away bird *C. concolor*. Native to much of southern and southeastern Africa (and often visiting suburban gardens and parks), it is indeed predominantly ashen in colour.

So too is the bare-faced or masked go-away bird *C. personatus*, but as its names suggest, its facial skin is bereft of feathers and is black in colour, yielding a mask-like appearance that contrasts sharply with its grey crest and the white plumes borne upon the remainder of its face and neck. This species has two discrete populations – one covering much of Tanzania, Rwanda, and Burundi as well as stretching into the peripheries of Kenya, Uganda, Zambia, Malawi, and the Democratic Congo, but the other confined entirely to Ethiopia.

Red-crested touraco *Tauraco erythrolophus*, 1838 painting

Happily, none of these three go-away bird species is endangered – much to the chagrin, no doubt, of the hunters all-too-frequently frustrated by their noisy scare-mongering!

Incidentally, the white-bellied go-away bird may have a claim to cryptozoological fame. It was nominated by W. Geoffrey Arnott in *Birds in the Ancient World from A to Z* (2012) as a possible identity for the pegasi - mythical horse-eared (or, according to Pliny the Elder, horse-headed) birds named after the legendary winged steed Pegasus. They reputedly lived by a lake in Ethiopia. The white-bellied go-away bird is indeed native to Ethiopia; and as noted by Arnott, it is known to perch in lakeside trees, and its crest recalls the shape of a horse's ear.

Visit http://www.youtube.com/watch?v=0N0ck5Uyf74 for an informative YouTube video featuring the curiously-named Mr McBouncyPants - a hand-reared white-bellied go-away bird at Houston Zoo, Texas.

Chapter 5:
CRYPTOZOOLOGY IN THE VATICAN

The collections of one of the world's greatest repositories of classical, medieval, and Renaissance culture[,] the Vatican Library, for six hundred years celebrated as a center of learning and a monument to the art of the book, is, nevertheless, little known to the general public, for admission to the library traditionally has been restricted to qualified scholars.

Introduction to *Treasures of the Vatican Library: And To Every Beast...*

A few years ago, a friend bought me a wonderful little book entitled *And To Every Beast...* (Millennium Books: Alexandria, NSW, Australia, 1994), which is one in a series of beautifully-illustrated thematic mini-volumes collectively entitled *Treasures of the Vatican Library*. All of them combine biblical quotations with illustrations selected from various tomes or manuscripts held in the vast collection of the Vatican's library (which contains over one million printed books, as well as 150,000 manuscripts and some 100,000 prints).

And To Every Beast... focuses upon animals, real and mythological, but it greatly intrigued me, because although it contains versions of several famous illustrations present in other bestiaries, it also includes some eye-opening pictures that I'd never seen before, of creatures that are so extraordinary as to be scarcely identifiable with anything known either to modern-day zoology or to zoomythology. This made all the more frustrating the fact that it does not state anywhere within its pages the original book or manuscript in the Vatican library that was the source of its pictures.

Happily, however, an online investigation via Google soon uncovered that elusive source publication. It was *The Animal Book*, written by famous Italian humanist and Renaissance author Pietro Candido Decembrio (1399-1477), commissioned by Ludovico Gonzaga, Marquis of Mantua, and published in 1460, with its illustrations added during the next century. So now, having solved that little mystery, here are some of this early tome's most fascinating if baffling illustrations, together with some commentary by me on what they may, or may not, depict.

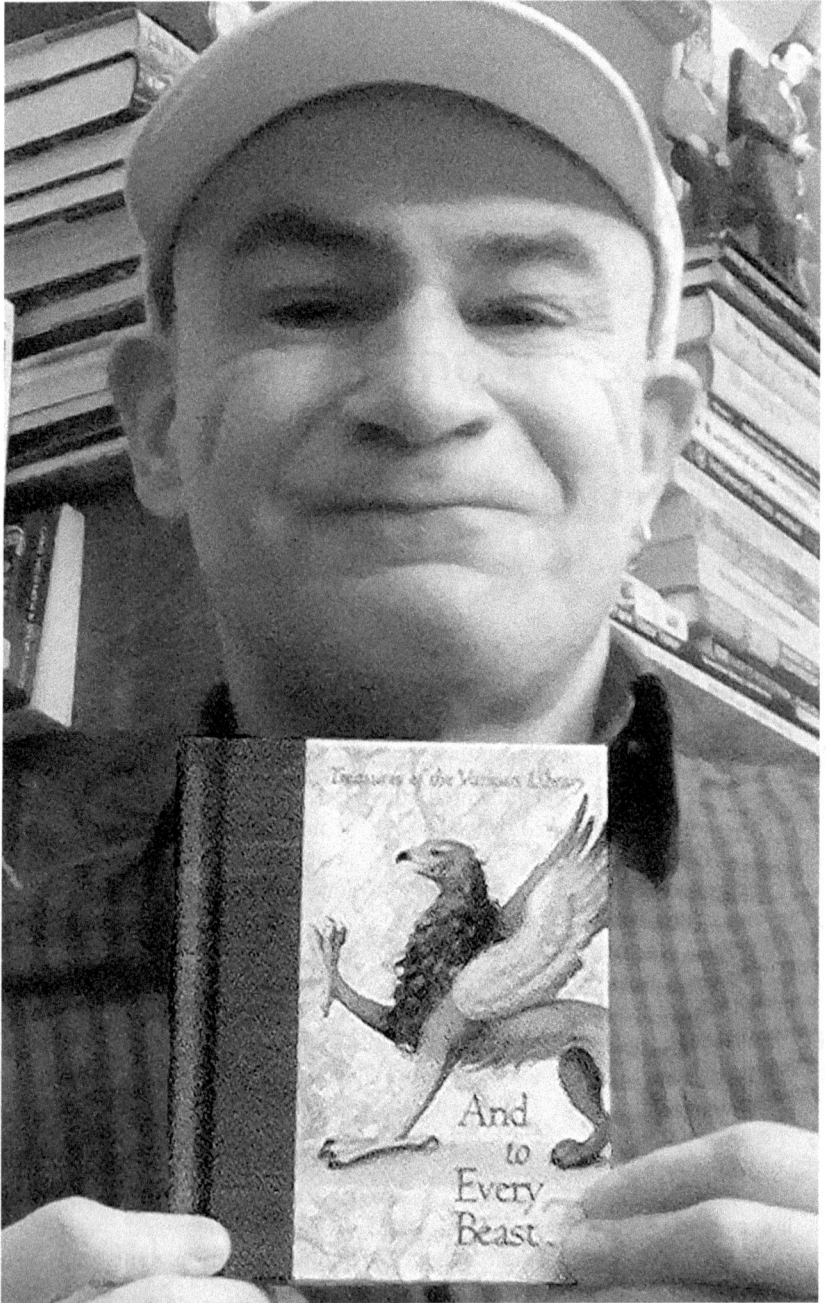

Holding my copy of *And To Every Beast...* (Dr Karl Shuker)

TRUNKO LIVES? NOT REALLY!
And where better to begin than with the totally bewildering, ostensibly aquatic mystery animal featured in the following illustration:

Scouring the Web, I have found that some sites have sought to identify it as a sea lion, but I see little if any resemblance to those particular pinnipeds. More plausible, even if only for the trunk-like proboscis, is an elephant seal *Mirounga* spp, but the depicted beast's long tail and, especially, its hoofed forelegs swiftly eliminate this from serious consideration. Unless, that is, the artist was attempting to illustrate such a creature merely from a verbal description (and quite probably a somewhat less than accurate one at that), rather than a physical specimen, because elephant seals weren't known scientifically until the mid-/late 1600s. Similarly, the aardvark *Orycteropus afer*, another postulated identity (albeit greatly distorted), remained scientifically undescribed until 1766; and the platypus *Ornithorhynchus anatinus*, which has also been suggested by some, remained unknown to Europeans until 1798.

When I first looked at this picture, the identity that came into my mind was that of a desman, especially the Russian desman *Desmana moschata* - that large aquatic relative of moles, which possesses a proboscis, a long tail, and clawed flipper-like hindfeet. However, this species' forefeet are also clawed and flipper-like, not hoofed (unless, once again, the artist was basing

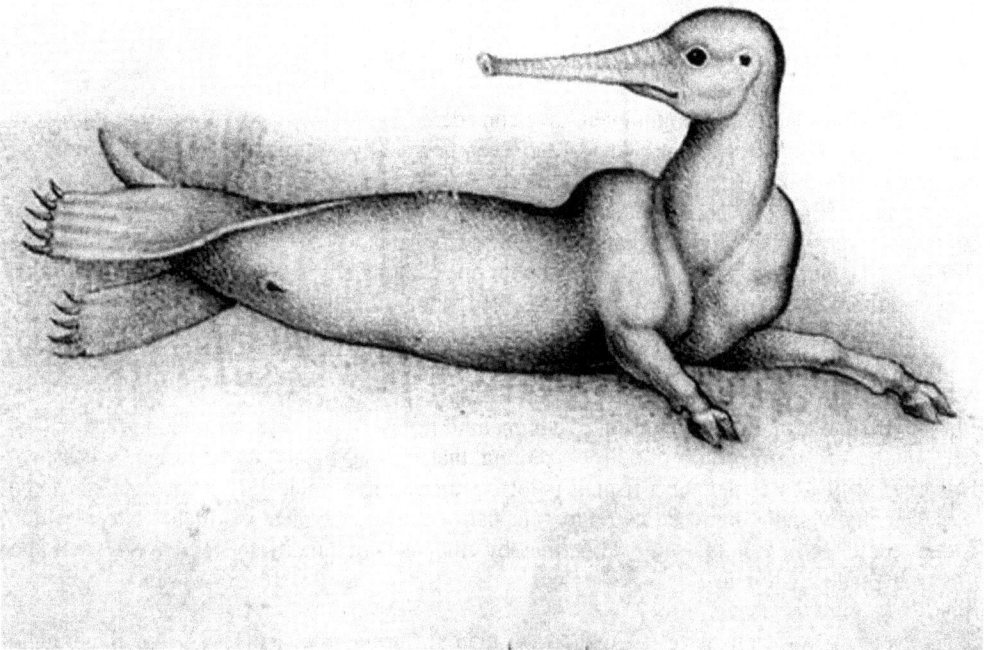

A hoofed, flippered, long-tailed, short-trunked mystery beast

19th-Century engraving of Russian desmans

his illustration upon an inaccurate verbal description only?).

Out of sheer desperation, I might even have considered the dramatic possibility that this was a portrait of the enigmatic Trunko - had I not been personally responsible for the latter entity's conclusive exposure as a non-living globster (see my book *Mirabilis*, 2013, for the full revelation). Perhaps the most reasonable assumption is that it represents some hoax taxiderm specimen, created via the skilful union of body parts from a variety of different creatures and displayed at sideshows or other public exhibitions, alongside stuffed mermaids, preserved dragons, dried Jenny Hanivers, and other assorted fauna of the fraudulent, fabricated kind.

A SCALY FISH-MAN?
Equally perplexing is this illustration of a humanoid figure completely covered in green scales. One might be forgiven for initially assuming that it was meant to represent a merman. However, it possesses neither a fishtail (sporting instead a normal, fully-formed pair of legs) nor even any webbing between its fingers, in stark contrast to typical mermen – one of which is depicted elsewhere in the same book, thereby emphasising the difference between itself and this weird scaly 'fish-man'.

Conversely, it may constitute an image of a man suffering from ichthyosis – a sometimes-extreme skin disorder in which the sufferer can indeed be covered in thick green scale-like flakes of skin. In the distant past, several unfortunate persons with this unsightly but striking

The scaly fish-man illustration

An oenophilic cynocephalus

medical condition have been displayed at freak shows and similar exhibitions, usually labelled as 'fish-men'.

OTHER ODDITIES
Much the same problems arise with many of the other creatures depicted in Candido's book, i.e. are they (a) mythical, are they (b) real but sometimes inaccurately represented, or are they (c) potentially cryptozoological?

Examples clearly belonging to the first of those three categories include a basilisk and a cockatrice, a griffin, an amphisbaena, a manticore, a flying horse, and a pair of onocentaurs; plus a donkey unicorn (or unicorn donkey?); a more typical white unicorn but with a red face and tricoloured horn; a manticore-related, tapir-reminiscent beast of legend known as a leucrocuta; a hairy dog-headed man or cynocephalus scrutinising a goblet of red wine; a peacock-crested, azure-breasted phoenix with crimson

An antelope-horned giraffe

wings and rooster-like wattles; and a bat-winged, scaly-bodied, limbless aerial dragon known as an amphiptere.

Those from the second category, meanwhile, include such familiar beasts as a lion, a lynx, a leopard, an eagle, a lammergeyer, an owl, an elephant, a stag, various hounds and horses, and a pelican, all depicted fairly or very naturally. Also present, but in stark contrast to the above-listed animals in terms of accurate representation, is a brown giraffe patterned only with a light speckling of small white flecks and armed with a pair of long curved antelope-like horns. Nor should (or could!) we overlook a gaudy multicoloured rhinoceros with extraordinary body armour that was clearly based upon the famous rhinoceros engraving of 1515 by Albrecht Dürer.

And creatures from the third (and most intriguing) category not only include the trunked mystery beast and scaly humanoid already discussed here. Also worthy of consideration are such curiosities as a snake bearing a veritable crown of horns around its brow; another one with a pair of lateral cow-like horns; a carnivorous deer (judging from its dentition) with antlers so palmate in shape that they resemble a pair of human hands; an even more mysterious carnivorous mammal with claws but a horse-like mane plus bright red fur mottled with grey and gold patches; and a frog sporting a pair of very long, slender, almost upright horns upon its head.

What could have inspired such zoological oddities? An exaggerated verbal description of a

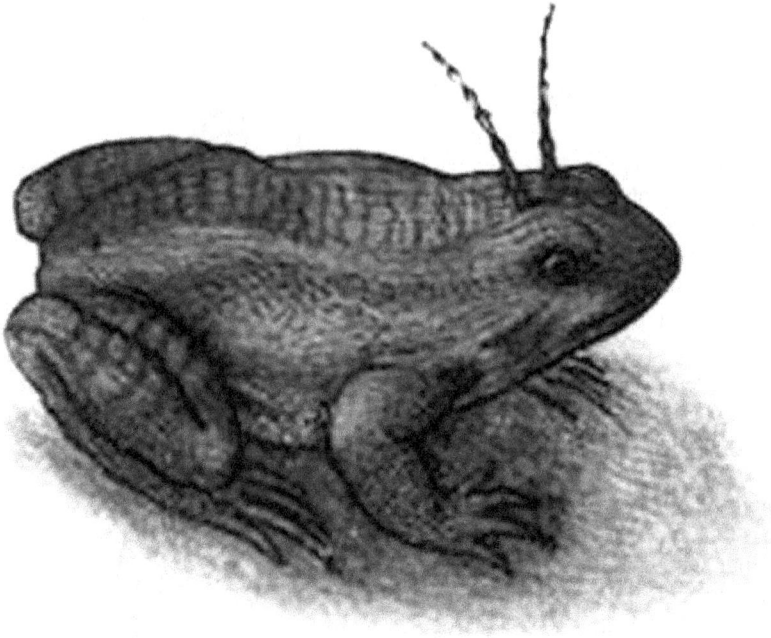

A frog with horns

horned viper may possibly have inspired the bovine-horned snake's depiction, but what of the others?

Whoever would have guessed that the Vatican was such a rich repository of cryptozoological and zoomythological exotica? And if you can obtain a copy of *And To Every Beast...*, please do so, because it's a thoroughly delightful, beautiful little book – highly recommended!!!

Chapter 6:
INTRODUCING THE NANDINIA
– AFRICA'S PARACHUTING (AND VERY
PARADOXICAL) EX-PALM CIVET

> Another scarcely-known little creature is the Nandinia...It is not a really savage animal: we once knew a tame Nandinia which had been captured when quite young in Africa and had subsequently accompanied her masters to France. She behaved as sweetly and gently as the most affectionate of cats, and she is no doubt still giving delight to all who know her. Nandinias stand on their hind legs from preference, doubtless so that they can see things from a greater height. This one, with her "arms crossed", as it were, displayed in this posture an astonishing grace and beauty. We live in an age where we care little about taming a new species of animal; yet there is no doubt at all that the Nandinia could become a really adorable companion.
>
> Robert Dallet and Robert Wolff – *Animals of Africa*

I've always been interested in the more unusual, enigmatic members of the animal kingdom, those that rarely attract such widespread attention as the giant pandas and great white sharks of this world.

And one creature that certainly embodies 'unusual' and 'enigmatic' is the nandinia *Nandinia binotata*, which was formally named and described by British zoologist John Edward Gray in 1830 (and originally split into two separate species - one West African and one East African). I first learnt of this cat-sized arboreal creature's existence when, as a child, my mother bought for me a huge, lavishly-illustrated book entitled *Animals of Africa*. Authored by Robert Wolff, it contained dozens of spectacular full-colour paintings by Robert Dallet, including one of the nandinia.

Even its name, 'nandinia', intrigued me, because it sounded so unlike any other animal name that I'd ever heard. Consequently, although I subsequently learnt that this species is also referred to as the African palm civet or two-spotted palm civet, as far as I'm concerned it will always be the nandinia. It is derived, incidentally, from 'nandinie' – a local West African name for this species.

The nandinia depicted upon an Ivory
Coast postage stamp issued in 1992

The nandinia depicted upon a Liberian postage
stamp issued in 1918.

The nandinia depicted upon a Togo postage stamp issued in 1965

Looking at its picture in *Animals of Africa*, the nandinia seemed to me to be a curious combination of cat and civet, with an attractive spotted coat and an exceptionally long tail, yet also indefinably different from anything else I'd ever seen. This proved to be quite a prescient assessment on my part, particularly at such a tender age, because although for many years the nandinia has been classed as a viverrid, i.e. a member of the taxonomic family housing the civets and genets, in recent times it has experienced a very dramatic recategorisation.

Genetic studies have suggested that during the evolution of the carnivorans (i.e. those species belonging to the mammalian taxonomic order Carnivora), the nandinia split off from the lineage leading to cats and viverrids *before* these latter two taxonomic families split from each other. As a result, it has been assigned to a taxonomic family of its own, Nandiniidae, of which it is the only member. As it is therefore an ex-palm civet now, this renders its alternative name of nandinia taxonomically unambiguous and thus much more desirable for use – another childhood impression of mine that has since proved prophetic.

But that was not the only surprise that the fascinating little nandinia had in store for me. While researching gliding vertebrates for a chapter in my book *Extraordinary Animals Worldwide* (1991), which reappeared in expanded form within its updated edition, *Extraordinary Animals Revisited* (2007), I accidentally discovered that the nandinia possessed a truly extraordinary but hitherto scarcely-publicised talent.

Resembling a portly, round-headed genet with very dense, woolly brown fur dappled with small black spots, and an extremely long, barred tail, but without any form of gliding membrane or other device for achieving aeronautical success, the nandinia is the most unexpected and least known of all mammals with gliding prowess. Indeed, I learnt of its surprising capability quite by accident myself. While flipping through some *African Wild Life* issues from 1958, I came upon a fascinating letter by G.V. Thorneycroft.

In his letter, Thorneycroft recalled seeing two nandinias high up in a tree one morning on his farm at Zomba, Nyasaland (now Malawi). One became frightened by his dogs, standing at the foot of the tree barking loudly, and as a result it chose to exit the tree in a quite astonishing manner. As noted by Thorneycroft, the nandinia:

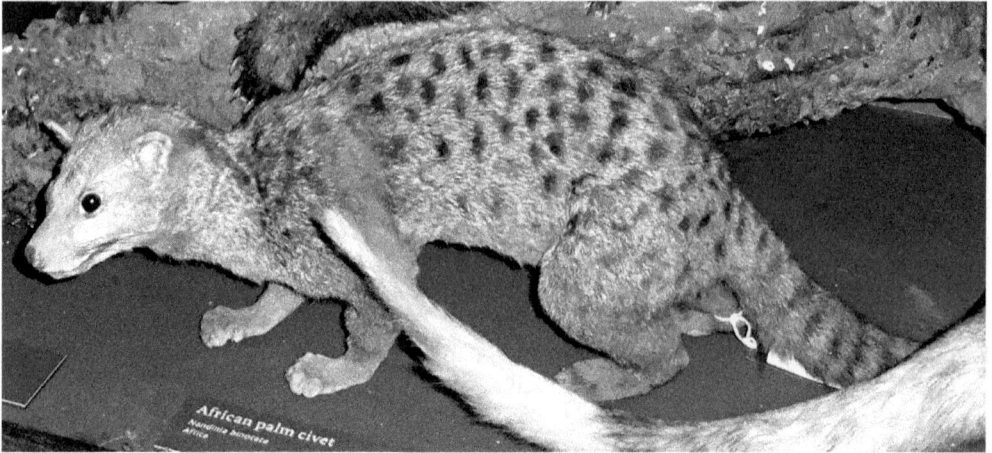

Taxiderm specimen of a nandinia at Tring Natural History Museum, Hertfordshire (Dr Karl Shuker)

> ...made a leap from a high branch and volplaned to the ground with legs and tail outstretched. It made a perfect landing on the bare ground, ran to another tree from which it again volplaned and repeated the action.

The mechanism responsible for this highly unexpected capability from as bulky and unlikely a gliding animal as the nandinia was revealed by Thorneycroft as follows:

> What struck me was the graceful way it planed or almost floated to the ground at an angle greater than half a right-angle so that it landed at a considerable distance from the tree it was in. Its tail was extended straight behind, the long hair at the base seeming to be 'on end' and its legs stretched out as far as possible. On each occasion it made a perfect four-point landing.

In short, the nandinia provided a surprising but nonetheless wholly corresponding verification of naturalist H.B. Cott's classic findings with gliding frogs back in 1926. Namely, that a fully outstretched body and limbs (with or without gliding membranes) is very important for successful aerial accomplishment.

I have never seen any further reports of parachuting ex-palm civets, but the nandinia has a very wide distribution across tropical Africa and has even been described as probably the most common small species of forest-dwelling carnivoran there. So unless this instance of volplaning was unique behaviour exhibiting only by the specimen observed by Thorneycroft, other occurrences must surely have been spied. Perhaps one day, therefore, some additional cases will be documented, substantiating the nandinia's claim to be the world's only species of temporarily-airborne carnivoran as opposed to being dismissed as a furry flight of fancy, in every sense.

An engraving of the nandinia from Rev. J.G. Wood's three-volume animal encyclopaedia *The Illustrated Natural History* (1859-63)

One further nandinian controversy does not seem to have been examined in any book, until now. Four nandinia subspecies are currently recognised (namely: *Nandinia binotata arborea*, *N. b. binotata*, *N. b. gerrardi*, and *N. b. intensa*), but in earlier days a much more mysterious form, nowadays relegated to a mere synonym of *Nandinia binotata*, was also recognised. This was *Paradoxurus hamiltonii*, Hamilton's palm civet, formally named and described by John Edward Gray in 1832. The only image of *P. hamiltonii* that I have seen is the following one, drawn by the eminent natural history artist Benjamin Waterhouse Hawkins in 1833-4:

Although a very elegant image, it does not appear to be a particularly accurate depiction of a nandinia (which is itself surprising for an artist of Hawkins's renown), but there may be good reason for this, inasmuch as the creature so depicted might not actually be a nandinia! Why I am so curious about it is the fact that the book in which Hawkins's painting originally appeared is the second volume of a collection of pictures owned by Thomas Hardwicke, entitled *Illustrations of Indian Zoology*.

Yet whereas all other palm civets are indeed Asian, the nandinia has always been an oddity among such civets zoogeographically, in that it is exclusively African. So if *P. hamiltonii* and *N. binotata* are truly nothing more than two different names for one and the same species, how can we explain the presence of an illustration of *P. hamiltonii* in a book devoted to the animals of India?

As far as I am aware, the aptly-named *Paradoxurus hamiltonii* remains a paradox today, but one that has long been forgotten. If anyone out there can shed any light on this

Benjamin Waterhouse Hawkins's illustration of *Paradoxurus hamiltonii*

anomaly, I'd be very happy to receive details.

Chapter 7:
UNEARTHING THE EARTH HOUND AND LABELLING THE LAVELLAN - SEEKING SCOTTISH MYSTERY MAMMALS

He held the trap and the dead animal up to examine it closer.

The animal was larger than a mole by far. It was as big as a good-sized rabbit. At first it might have been mistaken for an outsized rat but a closer inspection showed it was something quite different. Its head looked more like that of a small dog than a rat, but the large chisel-like incisors showed that it was *some* kind of rodent. Its front paws were spade shaped and equipped with impressive claws, obviously for digging. The hindquarters resembled a large rat, save for the bushy tail.

> Richard Freeman – 'See How They Run', in
> *Green, Unpleasant Land: Eighteen Tales of
> British Horror*

Among the characteristics ascribed to the animal [the lavellan] in various popular accounts are that it frequents water and damp places, that it strikes its victim with a discharge of venom, that no one in whose face it breathes will long survive, that it sucks blood like a vampire, that it has four feet, and that the length and breadth of its body are about equal, and measure from twelve to fifteen inches.

> *Transactions of the Gaelic Society of
> Inverness*, Series V, vol. 25 (1901-03)

Not all mystery beasts are huge monstrous creatures lurking in the seas, lakes, or remote rainforests of our planet. Some particularly intriguing examples are of much more modest proportions and live much closer to home, but are certainly no less interesting for that, such as the two small furry mystery beasts of Scotland examined here.

The earth hound envisaged as a mole-like cryptid (Shaun Histed-Todd)

THE EARTH HOUND OF BANFFSHIRE

In cryptozoological circles, Scotland is undoubtedly most famous for the Loch Ness monster, but a much less familiar creature of controversy is also on record from here - though as there do not appear to be any recent encounters with it, only unsubstantiated rumours, this very curious animal may no longer exist, always assuming of course that it ever did.

The Caledonian mini-beast that I am referring to is the earth hound of Banffshire, in northern Scotland. Also known as the yard pig, it reputedly builds simple nests in fields but preferentially inhabits or lives very near to graveyards, where it exhibits the decidedly unsavoury behaviour of digging inside any coffins that it finds in order to feed upon the corpses within them.

Reports of this macabre mammal are fairly sparse, with most dating back to the 19th and early 20th Centuries. One of the best descriptions was penned by a Mr A. Smith in an account from 1917. He described an encounter from half a century earlier (i.e. around 1867) between a gardener and an earth hound that the gardener had dug up while ploughing some haughs (alluvial flats) in Deveron close to a churchyard. The gardener described this mysterious animal as being brown in colour rather like a rat, but with a long hound-like head, and a tail bushier than a rat's.

After digging up the earth hound, the gardener swiftly killed it with the plough's swingletree (a horizontal wooden or metal bar) after it viciously bit through his leather boot. This creature's carcase was seen soon afterwards by a second eyewitness, who stated that it was about the size of a ferret, and looked midway in form between a rat and a weasel, but with a very dog-like head, and a tail of only modest length.

Additional morphological details were provided in an account of an earth hound killed in or around 1915 near Mastrick, Aberdeen, again close to a churchyard. Its observer claimed that it had mole-like feet (as might be expected from a fossorial animal), white tusks (fangs?), and prominent pig-like nostrils.

As 'earth hound' is a colloquial name sometimes applied to badgers, which are indeed known to invade churchyards, desecrate coffins, and devour corpses,

The earth hound, based upon eyewitness descriptions (William Rebsamen)

Artistic representation of the gardener's encounter with an earth hound (William Rebsamen)

one might assume that this is the explanation for Banffshire's enigmatic beast. However, even the most cursory reading of the above descriptions clearly reveals that the latter is fundamentally different in appearance from the familiar European badger *Meles meles*, instantly distinguished by its characteristic black-and-white striped face and much bigger size. Also, badgers construct extensive underground setts in forests, not simple nests in fields.

Another potential explanation that has incited some speculation is that the earth hound stories refer to young specimens of the wolverine *Gulo gulo* (adult wolverines can be up to 4 ft long and are the largest members of the mustelid family). Unfortunately, however, as with the badger suggestion, the morphology and lifestyle of the earth hound do not correspond at all with that of wolverines, of any age, which are not fossorial at all. In addition, whereas the badger is at least native to Britain, the wolverine is not, though it does occur in parts of northern mainland Europe.

Having said that, and as also documented in my book *Mysteries of Planet Earth* (1999), a few wolverine specimens have allegedly been sighted in modern times within various parts of Great Britain. If genuine, these may be escapees from fur farms (wolverines are rarely maintained in British zoos, though there are some currently on display at the Cotswold Wildlife Park). Even so, in every way the wolverine is simply too dissimilar from descriptions of the earth hound for this to be a viable identity.

Other identities that have been proposed are moles (but moles do not dig into coffins and eat corpses); rats (but rats do not possess digging feet, tusks, or hound-like heads, and brown rats do not build nests either); or ferrets (but ferrets fail to correspond with the earth hound for the same reasons as rats).

So could it be a species still-undescribed by science? Yet if such a distinctive beast as the earth hound had ever genuinely existed, surely there would be specimens of it in museums and other scientific collections? And yet no such examples exist – unless of course the grotesque necrophagous nature of its lifestyle warded off attempts to obtain and preserve specimens of this weird animal during earlier, superstition-ridden times when it still existed, and it has now simply died out?

Having said that: when an earth hound investigator called Alexander Fenton visited the Banffshire town of Reith in April 1990, he learnt that the earth hound is still spoken of there as a current creature of reality, not a mythical beast from bygone folklore. Fenton was even taken to a churchyard where such animals are said to dwell today, and which are described locally as between a rat and a rabbit in form. Fenton didn't find any there, but who knows? Perhaps the mystery of the Banffshire earth hound may yet be solved one day after all.

Incidentally: after reading my first, briefer coverage of the earth hound in my book *Mysteries of Planet Earth* (1999), cryptozoologist and horror fiction fan Richard Freeman was inspired to pen a suitably grim and grisly short story featuring this sinister cryptid. Entitled 'See How They Run', it appears in a recently-published collection of Richard's horror writings, *Green, Unpleasant Land: Eighteen Tales of British Horror* (2012). The earth hound is found to be a

19ᵗʰ-Century painting of some aquatic shrews

subterranean eusocial carnivorous mammal, existing in different morphological castes and headed by a single enormous queen – a scenario inspired by the real-life eusocial system of East Africa's naked mole-rat *Heterocephalus glaber*.

THE LAVELLAN OF CAITHNESS AND SUTHERLAND

A second small mystery mammal from northern Scotland is – or, again, was? – the lavellan. According to local lore in Caithness and Sutherland, apparently the stronghold of this cryptid, the lavellan was a rodent with flashing eyes, a disproportionately-large mouse-like or rat-like head, and similar body colouration too. However, it was larger than a rat, had an exceedingly venomous bite, was also a blood-sucker, and inhabited marshes as well as deep water-filled hollows in rivers.

Any cattle drinking from a body of water containing a lavellan would invariably die, and, bizarrely, this creature could inflict lethal injuries upon livestock from a distance too, from as far away in fact as approximately 100 ft, though the precise mechanism responsible for this fatal activity is never elucidated in such reports. Yet, paradoxically, if farmers had sick animals, they could be cured if they drank water in which the pelt from a dead lavellan had been dipped.

Interestingly, its name in Scottish Gaelic is also applied to the water shrew *Neomys fodiens* (which, interestingly, does have a weakly venomous bite) and the water vole *Arvicola amphibius*, both species having been identified as the lavellan by various authors. Yet the latter creature was supposedly much larger than either of them. Conversely, in John Fleming's book *History of British Animals* (1828), he claimed that it was likely to be the stoat *Mustela erminea*, because in early highland lore the stoat supposedly exuded some kind of "foul matter" that was toxic to horses and other animals.

The lavellan's most diligent modern-day investigator is naturalist Raymond Bell, who has memorably dubbed it a 'giant vampire shrew' in various talks and writings that he has prepared on this subject. He has speculated that it may have been at least in part nothing more than a fictitious bogey-beast invented by parents to ward their children away from deep water, or even an attempt to explain away mysterious diseases arising in livestock. However, he also concedes that some bona fide creature might have been at the core of the lavellan legend too, but what that creature was may never be determined.

It is fascinating to consider that the ostensible familiarity of Great Britain's extensively-studied, exhaustively-documented natural history can nevertheless still harbour such riddles as the earth hound and the lavellan. But will their mysteries ever be solved? Perhaps someone reading this chapter has the answer to that question and, if so, I very much look forward to hearing from you!

Engraving from 1840 of a hoopoe

Chapter 8:
HOOPOE, HOOPOE WHEREFORE ART THOU, BRIGHT BUTTERFLY BIRD OF MY YOUTH?

The Hoopoe fluttered forward; on his breast
There shone a symbol of the Spirit's Way
And on his head Truth's crown, a feathered spray.
Discerning, righteous and intelligent.
He spoke; 'My purposes are heaven sent;
I keep God's secrets, mundane and divine,
In proof of which behold the holy sign
Bismillah, etched forever on my beak.'

Farid Ud-Din Attar – *The Conference of Birds*

f I were asked to name my favourite species of bird, I'd have to give the matter considerable thought, bearing in mind that there are approximately 10,000 contenders alive and well and currently on record to choose from. If, conversely, I were asked to name my most exasperating but mesmerising species of bird, I could do so without any hesitation whatsover - *Upupa epops*, the hoopoe.

Named onomatopoeically after its triple-'hoop' cry, and resembling a gigantic pink butterfly with spectacular black and white wings, or an extravagantly ornamental Art Deco brooch designed by Erté and magically gifted with ethereal life, the hoopoe has fascinated me from my earliest days, ever since I first saw its elegant image gracing one of the pages of my now decidedly battered 1960s childhood copy of *The Observer's Book of Birds*. Reading its description, I was very excited to discover that this exotic species actually visits Britain annually, and has even bred occasionally in south England. Naturally, therefore, imbued with the eternal optimism that only a youngster can muster, I fervently hoped that one day soon my trusty Greenkat 10 x 50 binoculars would reward me with a sighting of this wonderful butterfly bird, possibly even within the urbanised surroundings of my West Midlands homeground.

As the years went by, I nurtured my hoopoe obsession by reading whatever I could find concerning this enigmatic creature. And so I learnt all about its predominantly insectivorous diet and the surprising oil-ejecting defence behaviour of its chicks; was shocked by lurid details of its disgustingly filthy nests and its vicious territorial battles; was startled by the unexpected description in 1975 of subfossil remains from a hitherto-unknown species of giant hoopoe *Upupa antaios* on the South Atlantic island of St Helena; and in particular was very enamoured by the wealth of folklore and legends associated with this feathered icon.

According to one ancient Arabian tradition, for example, hoopoes originally bore crests of solid gold, bestowed upon them by King Solomon in gratitude for shielding him with their wings from the burning sun one day as he walked through the desert. So many of their number were killed for this valuable accoutrement, however, that eventually they came before Solomon, who was so wise that he could even understand the language of birds, and beseeched him to help them. Touched by their tragic plight, Solomon agreed to do so, as a result of which the hoopoes' crests were transformed from gold into feathers, thus saving their species from extinction.

The hoopoes are also said to have brought to Solomon the shamir (see Chapter 21 for a detailed coverage of this biblical cryptid) – described in the Talmud and Midrash as a tiny but very magical worm that could cut through solid stone, and which greatly assisted him, therefore, in building his First Temple in Jerusalem. (In a similar vein, the hoopoe is also credited with knowledge of where to find a mystical plant called the springwort, whose touch can break through the hardest rocks and stones.) And in the Koran, it was the hoopoe that discovered the Queen of Sheba and informed Solomon of her existence. Other Arab traditions claim that the hoopoe could unerringly guide Solomon to undiscovered subterranean springs by using its long bill as a water-divining rod, and consider it to be a doctor among birds, gifted with medicinal powers that can cure any ailment.

Meanwhile, in Greek mythology, King Tereus of Thrace, his queen Procne, and her young sister Philomena became embroiled in such a hideous saga of rape and bloodshed that Zeus transformed all three of them into birds. Tereus became a hoopoe, Procne a swallow, and Philomena a nightingale.

By the time of my late teens, I realised, sadly, that just like so many other dreams, mine of seeing a hoopoe in England was probably destined never to be achieved. Consequently, if the hoopoe would not come to me, I would have to go to it. And so it was that during the summer of 1978 I went on a coach-touring holiday in Andalucia, Spain, where I hoped to espy not only the hoopoe but also two of its cousins.

The taxonomic order Coraciiformes contains some of the most colourful families of near-passerine birds, in particular the kingfishers, the rollers, the bee-eaters, and the hoopoe/wood hoopoes (nowadays the two hoopoe groups are generally split into separate families but back in the late 1970s they were still combined). Moreover, at least one species from each of these families could be found in Andalucia. Namely, the European kingfisher *Alcedo atthis* (which I'd already seen in Britain), the European roller *Coracias coracias* (a very rare summer

Painting of a pair of hoopoes by John Gould, from 1837

19th-Century engraving of a hoopoe

vagrant in Britain), the European bee-eater *Merops apiaster* (a rare summer visitor to Britain and very occasional breeder here), and of course the hoopoe itself.

Sadly, the roller never made an appearance, but bee-eaters were readily visible perching on telegraph wires, as pointed out to me by our tour guide, Pedro - who, as good luck would have it, was also an enthusiastic amateur ornithologist, always carrying a birdwatching field guide and binoculars with him. Consequently, he ensured that I never missed any of his region's native avifauna during our various trips. One evening, moreover, he informed me to my great delight that during our sightseeing tour the next day we would be in an area where hoopoes were commonly sighted! Finally, at the age of 18, I would be seeing my long-awaited butterfly bird – the best coming-of-age present that I could have wished for!

Fate, however, had other ideas. During that same night, I fell ill with an acute stomach bug, which was so severe that I had no option but to forego my planned tour the following day and spend the whole time in bed instead. That evening, when my party returned from the tour, Pedro came to see how I was, and informed me that they had indeed seen hoopoes – which, if anything, made me feel even worse than the stomach bug had succeeded in doing! Never mind, I consoled myself, surely I would see some before too long? And indeed I did see some – the only problem was that in my case the period "before too long" turned out to be 30 years!

Had good fortune shone upon me, however, it might not have been quite so long, and I wouldn't even have needed to travel very far in order to fulfil my undiminished ambition of seeing a hoopoe.

It was late afternoon on Monday 9 October 2006, and I was casually flicking through my local evening newspaper, the *Express and Star*, when I suddenly spotted a report stating that during the weekend just gone, dozens of birdwatchers from all over the country had descended upon the grounds of a closed-down local school - because, totally unexpectedly, a hoopoe had appeared there! The report even included a photo of the bird, which was unquestionably a hoopoe. Moreover, situated in the West Midlands town of Walsall, the derelict school in question, Beechdale Primary, was only a few miles from where I live! If only I'd known about this visitation earlier!

Nevertheless, as soon as I'd read it all, I cut the report out of the newspaper, stuffed it in a back pocket of my jeans for further reference if needed, grabbed my binoculars, jumped on my motorbike, and rode off straight away to the school, in the fervent hope that my elusive butterfly bird would still be there and show itself to me.

Needless to say, of course, the hoopoe did no such thing. After over an hour of training my binoculars on every blade of grass, bush, branch, and twig in the vicinity, I gave up in total despair. Clearly, to quote an old but very apt maxim, the bird had flown.

And so I had no option but to ride back home, frustrated and thoroughly dejected - tormented yet again by this feathered phantom that I seemed destined never to see, not even when it was almost in my own back garden!

Fantasy hoopoe depicted on a mirror purchased by me several years ago (Dr Karl Shuker)

Instead, I would have to content myself with purchasing at an absolute bargain price a very large, attractive mirror depicting a gorgeous fantasy hoopoe that I happened to spot one morning in the window of a local charity shop; and also with continuing to uncover interesting if somewhat esoteric snippets of hoopoe folklore and legend from around the globe - because that at least appeared to be something related to this infuriating entity at which I was able to achieve a modicum of success.

I discovered, for instance, that many cultures throughout its extensive Eurasian and African zoogeographical distribution range traditionally deem the hoopoe to be a guide or leader of other birds through dangerous realms to their ultimate destination, as well as a messenger from the invisible supernatural world (this latter role of the hoopoe also features in Aristophanes's famous play, *The Birds*), and a keeper of God's secrets. To the ancient Egyptians, it symbolised gratitude, and even appeared as a hieroglyphic. There is also a widespread folk tradition that the hoopoe can forecast storms. Bearing in mind, however, that scientists have shown that it can indeed detect minute atmospheric electrical (piezoelectric) charges that sometimes precede a storm or even an earthquake, this particular example of hoopoe folklore is clearly based upon fact.

In addition, the hoopoe was viewed as a harbinger of war in Scandinavian legends, and associated in Estonian lore with death and the underworld. Acquiring a more positive role, conversely, in May 2008 it was chosen as the national bird of Israel, and is also the state bird of India's Punjab province.

I was even able to solve a hoopoe-related mystery that had baffled me for many years. As a child, I was given as a gift a large and beautifully-illustrated book appropriately entitled *The Colourful World of Birds*, published in 1963 and written by acclaimed French ornithologist Jean Dorst (see also Chapter 2). In a spread on extinct birds, a subject that had always interested me, Dorst briefly name-checked a species that had vanished during the 19th Century but which

Hoopoe-seeking by Harley! (Dr Karl Shuker)

was totally unfamiliar to me – the 'Bourbon Island hoopoe'. And despite my diligent perusing through numerous ornithological tomes during subsequent years, I never encountered any further mention by that name of this mysterious lost bird. Eventually, however, I discovered that 'Bourbon Island' was an old name for what is nowadays referred to as the island of Réunion, a neighbour of Mauritius in the Indian Ocean's Mascarenes group. As for its hoopoe, this proved to be a mistranslation of 'huppe', the local name for a now-demised but very extraordinary species of starling.

Known scientifically as *Fregilupus varius*, the Réunion crested starling, its slender curved bill (more curved in females than in males) and distinctive but very unexpected crest (for a starling) bestowed upon it a surprising resemblance to a hoopoe, hence its local name. Indeed, when first described scientifically in 1783 by Dutch naturalist Pieter Boddaert, this aberrant starling was actually thought to be a species of hoopoe and so he christened it *Upupa varia*; only later was its identity as a starling established and its generic name changed accordingly.

Réunion crested starling, painted by J.G. Keulemans, 1907

I had always promised myself that one day I would go on safari. And so it was that in November 2008 I found myself staying at the private Shamwari Game Reserve in South Africa's Cape Province, not far from Port Elizabeth. During what remains the best holiday of my life, I was able to observe an unparalleled abundance of wildlife – everything from lions, cheetahs, warthogs, hippopotamuses, buffaloes, giraffes, baboons, rhinoceroses, zebras, ostriches, and a vast diversity of antelopes to such rarer, more elusive species as servals, brown hyaenas, springhaases, mongooses, caracals, stone curlews, nightjars, and even two different leopard specimens (many visitors don't even manage to catch sight of one). But for me, the greatest highlight of all happened entirely without warning.

On the morning of 4 November, while walking through the reserve's gardens towards the jeep to get aboard for the first game drive of that day, what looked like two gargantuan cerise butterflies flapped by overhead. Training my binoculars upon their undulating flight, I froze as if petrified by Medusa herself, for as they alighted upon a branch of the tree nearest to the jeep, their pied crests matching their eyecatching wings, I realised that what I had just seen was a pair of hoopoes!

There at last, before my unbelieving eyes, was my spellbinding, evanescent butterfly bird, and suddenly three long decades of disappointment simply melted away. Just for a moment, I was an 18-year-old youth again, but what had then been nothing more than the excitement of

Public domain photograph of a hoopoe in the wild

anticipation was now replaced by the thrill of fulfilment. Later on during that same South African holiday, I saw hoopoes again, and I even spied a close relative – an exquisite green wood hoopoe *Phoeniculus purpureus*, with slender coral-red bill, and richly garbed in gorgeous viridescent and metallic purple plumage.

Yet nothing could surpass that first morning encounter, when at long last my eyes were blessed by the sight of what may not have been the sweet Bird of Youth but which was in many ways the sweet bird of *my* youth.

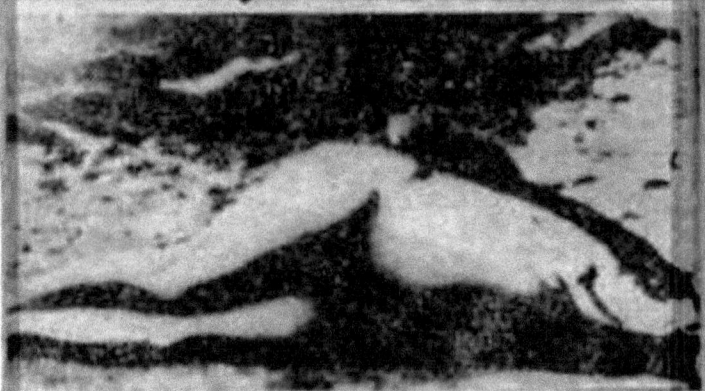

Some fishy news from Cairo

HERE'S the dishiest fish you've ever seen—a mermaid with a fish's head and the hips and legs of a fully-developed woman. A Cairo newspaper claims it was caught in the Red Sea off Yemen. If one ever gets to Ireland it would hot up the Cod War . . .

Unidentified newspaper cutting of 7 April 1973 documenting
the reverse mermaid of Cairo

Chapter 9:
RENÉ MAGRITTE AND THE REVERSE MERMAID - A VERY FISHY TALE, IN EVERY SENSE!

SOME FISHY NEWS FROM CAIRO
Here's the dishiest fish you've ever seen – a "mermaid" with a fish's head and the hips and legs of a fully-developed woman. A Cairo newspaper claims it was caught in the Red Sea off Yemen. If one ever gets to Iceland it would hot up the Cod War.

Unidentified British tabloid newspaper article, 7 April 1973

My cryptozoological archives contain many strange reports, but few are stranger – or fishier – than the extraordinary case of René Magritte and the 'reverse mermaid' from the Red Sea.

Even as a child, I would always cut out any unusual animal-related reports that I'd spotted in the various newspapers that my parents bought each day, and would then diligently paste or sellotape them into a series of large scrapbooks, which I still have today. I would even jot down alongside each report the date on which it had been published, but, sadly, my youthful zeal did not always extend to writing down the name of the newspaper in which it had been published.

So it was that on Saturday 7 April 1973, I spotted a brief but truly bizarre article in one of the several London newspapers that my parents had purchased that day, and promptly cut it out, but without noting down the newspaper in question. Looking back, however, I am certain that it was a tabloid newspaper – most probably either the *Daily Mirror* or the *Sun*. After all, I could hardly imagine the *Times* or *Daily Telegraph*, for instance, publishing the article reproduced on page 84 opposite.

As seen, the article included an extraordinary photograph of what can indeed be described as a mermaid, but, uniquely, a mermaid in reverse of the usual type, inasmuch as it combined the upper half of a fish with the lower half (legs) of a woman, instead of the other way round.

Needless to say, at the age of 13 I was sufficiently worldly to recognise that this bizarre entity was evidently a fake, but a fascinating one nonetheless, and so this intriguing clipping was dated and sellotaped into one of my scrapbooks. There it remained for a number of years until my interest in cryptozoology had flourished to the point where I had begun amassing files of reports, scientific papers, and other publications as a discrete cryptozoological archive. Accordingly, the tabloid report of Yemen's reverse mermaid was carefully removed from its scrapbook home and rehoused in a file exclusively devoted to merfolk reports, where it remains to this day, constituting a truly exceptional account.

During the 1990s, while corresponding with fellow cryptozoological enthusiast Michael Playfair regarding a number of different mystery beasts, I mentioned my reverse mermaid report to him, and as he expressed interest in it I photocopied it and posted the copy to him. Like me, Mike was very intrigued by the weird, reverse nature of the entity's composite form, so he promised to investigate the case and let me know if he uncovered anything relevant.

True to his word, a few weeks later Mike posted me a photocopied page from an encyclopedia of science-fiction movies, the page in question reproducing a film still that depicted a surprisingly similar entity to the version photographed in my newspaper cutting. The still was from an American science-fiction feature movie entitled *Phase IV*, which had been made in 1974 (i.e. just a year after the newspaper report had been published), and it had won the Grand Prix at the 1975 Trieste festival of science-fiction films.

The only feature movie ever made by Saul Bass, a noted designer of film titles, *Phase IV* starred Michael Murphy, Nigel Davenport, and Lynne Frederick, and was shot in England's Pinewood Studios (even though it was actually set in the United States). It told the story of how, following some undescribed multi-phase cosmic event, a colony of ants in the Arizona desert had undergone a form of accelerated evolution culminating in the development of a collective or 'hive' mind, and had become antagonistic towards a scientific team sent out to investigate the extraordinary geometric patterns and towers that the ants had been creating there. Davenport played a typical 'mad scientist' character, whose experiments in evolution had created, among other things, the reverse mermaid in the film still.

All very interesting, but unfortunately it did not provide an explanation of my newspaper report, because that had been published a year before *Phase IV* had even been released. Mike was unable to discover anything further of possible relevance, and for quite some time I was unable to do so either – until one day, that is, when the Belgian surrealist artist René Magritte (1898-1967) came to our rescue.

I had long been captivated by Magritte's extraordinary art, ever since, in fact, while still a child, I had encountered in a set of encyclopedias his fascinating painting 'The Voice of the Winds' - depicting a trio of enormous alienesque spheres floating with somewhat

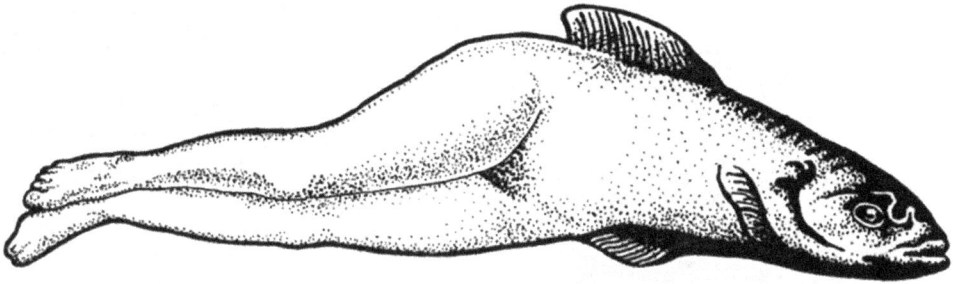

Line-drawing representation of the reverse mermaid in Magritte's painting 'Collective Invention'

menacing presence above a field. Now, many years later, I had been idly flicking through a book of Magritte's paintings in a book fair – when, abruptly, I came upon one that I had never seen before and which quite literally took my breath away.

For although I had not previously seen the painting (which was entitled 'Collective Invention' and had been produced by Magritte in 1935), its image was instantly familiar to me. It was a reverse mermaid, but not just any reverse mermaid. Even though I hadn't looked at the newspaper report from my scrapbook for a long time, it seemed to me that the entity in Magritte's painting was not merely similar – but was identical – to the reverse mermaid in the report.

Purchasing the book, as soon as I was back home I checked the reverse mermaid in 'Collective Invention' with that in the newspaper report, and confirmed that they were indeed one and the same, differing only in that the shadow effect on the entity in the painting had been enhanced in the report's version to render it more life-like. The mystery of the reverse mermaid from the Red Sea was no more – it was simply a curious fraud perpetrated by some still-unknown hoaxer(s) and inspired by one of Magritte's surreal works of art.

In early March 2011, I posted a photograph of 'Collective Invention' on my Facebook wall, and in response FB friend Igor Burtsev provided the following information, which independently corroborated the time period and content of my old newspaper cutting:

> I was in Aden at that time, 1972-73, when local youth had shown this pic as the photo of a real beast. The people there believed in its reality! I couldn't disillude them of that!

Clearly, therefore, the hoax had indeed originated in or around Yemen during the early 1970s. All in all, a very fishy affair in every sense, but one that, unlike so many others in the cryptozoological world, has finally been at least partly resolved. And who knows? Perhaps one day the still-undisclosed story and perpetrator(s) behind the hoax itself will come to light, and this decidedly odd pseudo-cryptozoological case can then at last be closed.

Chapter 10:
THE PICHICIEGO
– ARGENTINA'S ARMOURED FAIRY

> In the course of a day's ride, near Bahia Blanca, several [pichiciegos] were generally met with. The instant one was perceived, it was necessary, in order to catch it, almost to tumble off one's horse; for in soft soil the animal burrowed so quickly, that its hinder quarters would almost disappear before one could alight.
>
> Charles Darwin – *The Voyage of the Beagle*

A cryptozoological animal is usually defined as one whose existence is known to the local people sharing its habitat but which is not recognised by science. All of the major new terrestrial species scientifically described and named since 1900 fitted this definition prior to their official discovery and recognition – the okapi, mountain gorilla, giant forest hog, Komodo dragon, kouprey, Congo peacock, Chacoan peccary, Vu Quang ox, giant muntjac, and kabomani tapir, to name but a few.

Only one exception to the rule is generally named – the Congolese water genet (aka civet) *Osbornictis piscivora*. For according to Dr Bernard Heuvelmans's classic cryptozoological tome *On the Track of Unknown Animals* (1958), as well as many other sources too, this extremely reclusive mammal was wholly unknown to the local people when its holotype was obtained in a forest stream at Niapu, in what is now the Democratic Congo, on 1 December 1913 by two American Museum of Natural History scientists – Dr James Chapin and Herbert Lang. In reality, however, as I first revealed in my book *The Lost Ark: New and Rediscovered Animals of the 20th Century* (1993), this oft-retold tale is a fallacy. When that first specimen was procured and shown to native hunters in Niapu, they did recognise its species, and in the local Kibila and Kipakombe languages it even had its own specific name – the esele.

Notwithstanding this, there is at least one very distinctive mammal whose existence had genuinely remained hidden from its human neighbours until its formal scientific discovery – and that is the pichiciego (aka pichiciago) or pink fairy armadillo. No bigger than a mole, this extraordinary little creature is the world's smallest species of armadillo, and is native to the dry grasslands and sandy plains of central-western Argentina. For much of the time it remains underground, burrowing in soft sand with incredible alacrity, and using its huge front claws to agitate the sand so that it can then quite literally swim through it, as easily as if it were swimming through water.

So rarely is this cryptic animal seen above-ground, however, that the local people were astonished when in 1824 zoologist Prof. Richard Harlan from the University of Philadelphia unearthed a pichiciego at Mendoza. Nothing like it had ever been seen before, either by them or by zoologists, and when he officially described its species in 1825, Harlan christened it *Chlamyphorus truncatus* – sole member of an entirely new genus. And once the local people became more familiar with this minute armadillo, they dubbed it 'pichiciego' - a name derived from Mapuche, which is an indigenous language spoken in part of this species' zoogeographical range. In Mapuche, 'pichi' means 'small', and 'ciego' means 'blind' – so 'pichiciego' translates as 'little blind one'.

Unlike the much thicker, tougher, and more comprehensive body armour of other armadillo species, the plates of the pichiciego are notably thinner, and only loosely attached to its skin by a thin membrane running along the vertebral column. In addition, none but its pelvic armour is hardened, and breast armour is completely lacking. However, its head armour extends beyond the horny nasal plate (probably to compensate for the fragility of the creature's skull), and its broad-ended tail also has a covering of armour. As one might expect from a predominantly subterranean, fossorial species, the pichiciego's eyes are extremely small, and its external ears are nothing more than tiny folds of skin.

Even today, this most minuscule of armadillos is so elusive and seldom seen that television naturalist Nick Baker's quite recent search for it in Mendoza, as featured in an entertaining episode of his *Nick Baker's Weird Creatures* BBC series from 2008, ended in dismal and seemingly inevitable failure. Despite setting 20 pitfall traps and persuading the local fire

Engraving from the 1840s of the pichiciego

19ᵗʰ-Century engraving of a pair of pichiciegos

brigade to drench the search area's sand dunes with water in the hope of stimulating some pichiciegos to surface (after learning from the locals that these rarely-spied creatures appear more frequently after rainfall), a dejected Nick ultimately conceded defeat. He had to be satisfied instead with viewing a preserved specimen in the home of a villager. Indeed, on the few occasions when a pichiciego is encountered, it is usually captured alive and maintained in the living state for as long as possible by its fascinated captors before it is finally preserved in mummified form as a highly-prized curio.

But perhaps Nick and others have simply looked for pichiciegos in the wrong way or even in the wrong place. During his own search for this evanescent mini-mammal in Mendoza's pampas region, 20ᵗʰ-Century German zoologist Wolf Herre journeyed to an isolated railroad station in the middle of a broad, flat expanse of pampas that was said to be the pichiciego's centre of distribution – even the area's shrubbery was named 'pichiciego' after it. And here he eventually did achieve success, albeit in a most surprising way:

> For hours we wandered under the bright sun, trying in vain to find tracks of the pichiciego. But in the vicinity, wooden railway ties were being replaced, and the pichiciego was found in the rotting wood, probably attracted by the insect larvae inside.

Altogether a very remarkable creature – so remarkable, in fact, that in the years following the pichiciego's discovery by Harlan, no one expected to find anything else even remotely similar.

19ᵗʰ-Century engraving of pichiciegos

In 1859, however, while crossing the Andes via a route not previously utilised by Western travellers, German zoologist Prof. Karl Hermann Burmeister discovered in the possession of a native Indian villager the mummified remains of a hitherto unknown species of armadillo that superficially resembled a larger version of the pichiciego, and thus became known as the greater pichiciego.

This intriguing species' generic name has changed several times since it was scientifically described by Burmeister in 1863, and some zoologists no longer consider it to be closely related to the pichiciego. However, it is still commonly referred to as the greater pichiciego, and is known scientifically as *Calyptophractus retusus*, the only member of its genus. Native to Argentina, Paraguay, and Bolivia, even today it remains scarcely known. Indeed, much of the little information that has been documented about this mysterious mammal was gained from observations made of a living specimen captured in Mendoza during July 1965, which was subsequently sent to Chicago's Brookfield Zoo, where it lived until its death in December 1971.

Finally: the pichiciego has very rarely been filmed alive, but I have tracked down three short YouTube videos of living specimens. These are, respectively, as follows:

http://www.youtube.com/watch?v=cePw_kK1Wdk

http://www.youtube.com/watch?v=07Sor7uPtK4

http://www.youtube.com/watch?v=n_TmLA4h9FM

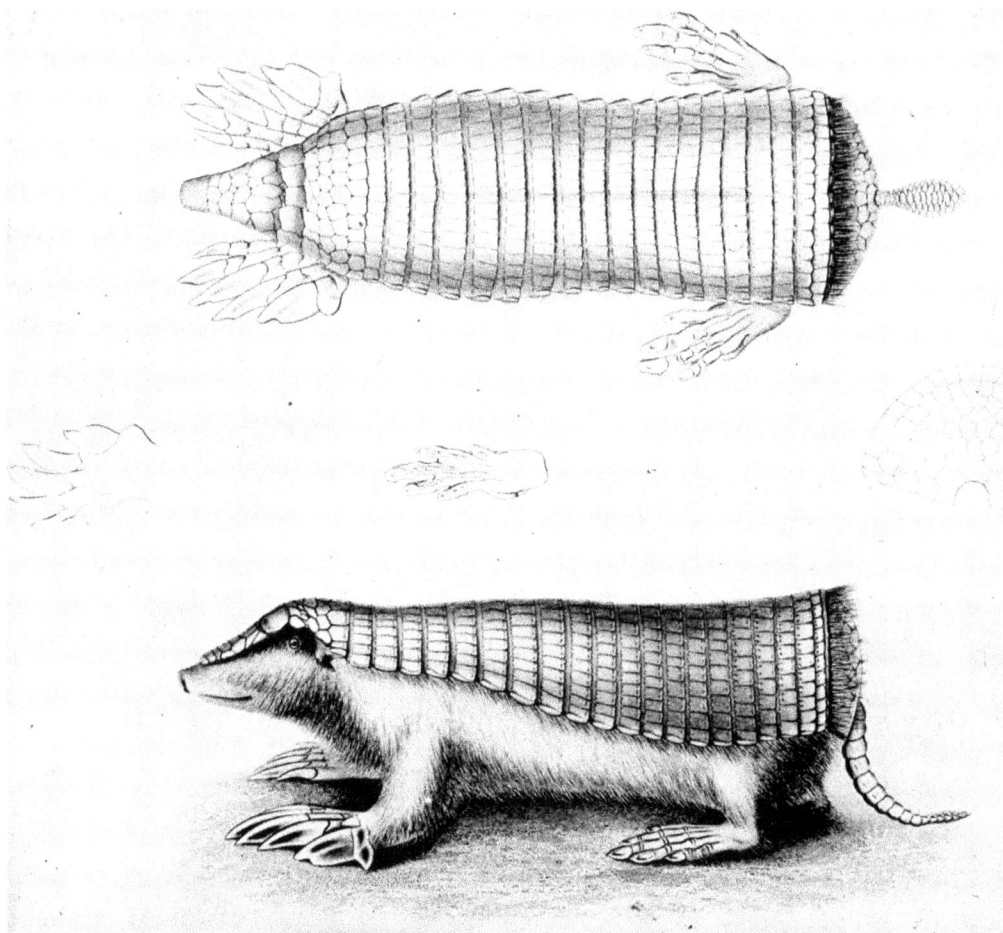

1855 engraving of the pichiciego – dorsal and lateral views

Not only do they provide a seldom-seen insight into its unexpectedly rapid locomotion above-ground, its spectacular burrowing abilities, and its ridiculously tiny size, but also they readily demonstrate why the pichiciego or fairy armadillo is unquestionably one of the most enchanting little animals on our planet.

Chapter 11:
VANISHED RATS AND MISSING MARMOTS - RODENT MYSTERIES SOLVED BY LADY LUCK

[The] island we found uninhabited, and it contained plenty of trees, and so many birds, both marine and land, that they were without number...and we saw no other animals except very big rats and lizards with two tails, and some snakes.

Amerigo Vespucci(?) - *Lettera di Amerigo Vespucci delle Isole Nuovamente in Quattro Suoi Viaggi*

The following mystery rodents were lost and forgotten for a very long period of time, only to be refound and their respective identities finally resolved much more recently - and in each case assisted in no small way by the random favours of Lady Luck - as now revealed here.

THE VANISHED RAT OF VESPUCCI

The Italian explorer Amerigo Vespucci (1454-1512), after whom America is popularly believed to have been named, allegedly made four separate voyages to the New World, the first two leaving from Spain and the second two from Portugal. However, there is some dispute among historians as to whether the first and fourth of these voyages did actually take place.

The only source of information for the fourth one is a published document known in full as the *Lettera di Amerigo Vespucci delle Isole Nuovamente in Quattro Suoi Viaggi*. Printed in 1504 or 1505, it was purportedly written by Vespucci to prominent Italian statesman Piero Soderini (1450-1522) of Florence, and in it he described all four voyages. However, not everyone is convinced that it truly was Vespucci who authored this document. Yet even if it wasn't, the descriptions of the Brazilian locations and wildlife contained in it are so accurate that they were clearly written by someone who had genuinely visited and observed them.

Vespucci's supposed fourth voyage consisted of the sea journey from Portugal to the New World and then down Brazil's Atlantic coast before returning to Portugal, the entire voyage lasting from May 1503 to June 1504. One section of the letter's account of this voyage is devoted to a small volcanic island not hitherto visited by Western explorers. It became known as Fernando de Noronha, is situated off Brazil's easternmost point, and among the fauna encountered there were some mysterious rodents ostensibly seen by Vespucci when visiting there on 10 August 1503 and referred to by him as "very big rats" (see quote opening this present chapter). Bearing in mind that no Westerners had previously visited this island, these rodents must therefore have represented a native species, rather than simply being European black (ship) rats *Rattus rattus* introduced onto it from Western ships.

Engraving from 1896 of black rats

Having said that, however, following subsequent visits to this island by other Western ships, European black rats did indeed eventually find their way ashore there, and soon populated the island en masse. By the time that it was visited by biologist H.N. Ridley in 1887 after having been virtually ignored by science since the time of Vespucci, these rats were everywhere, but there was no sign whatsoever of the larger, apparently native version referred to in Vespucci's letter. So even if such a creature had ever existed, it was clearly long-extinct, no doubt out-competed and exterminated by the invading European black rats, which may have preyed upon this native rodent's young and carried pathogens lethal to it.

But had it ever existed to begin with, or was what became known as Vespucci's rat just an invention by the letter's writer, regardless of the latter person's identity? After all, there was no physical evidence whatsoever to substantiate this animal's erstwhile reality – until August 1973, that is.

This was when Smithsonian Institution ornithologist Dr Storrs L. Olson, seeking bird fossils at the eastern corner of Fernando de Noronha, inadvertently uncovered in some old beach dunes various late Holocene remains of a sturdy rat-like rodent larger than Europe's familiar black rat, and estimated to have weighed as much as 7.0-8.8 oz (black rats typically weigh only around 4.6 oz). Olson and Smithsonian Institution mammalogist Dr Michael D. Carleton duly subjected these intriguing remains to a detailed morphological study that confirmed their originator to be not only a new species but also sufficiently distinct from all other known rodents to require the creation for it of an entirely new genus.

Amerigo Vespucci statue in Florence, Italy

In March 1999, Carleton and Olson co-authored their very novel rodent's official scientific description, published in the journal *American Museum Novitates*. And mindful of the mystery of those perplexing "very large rats" allegedly sighted on this same island by Vespucci, they christened their disinterred rodent *Noronhomys vespuccii*.

In so doing, they acknowledged that this sizeable species (shown to be most closely related to the rice rats *Oryzomys* spp.) probably still existed there as late in time as the 16th Century, and thus would have been what Vespucci (or whoever wrote the letter to Soderini) had seen and referred to. Thanks, therefore, to a fortuitous discovery of its remains, Vespucci's vanished rat was a mystery no longer.

A rice rat, specifically the marsh rice rat *Oryzomys palustris*

THE QUEMI QUESTION
The hutias constitute a series of coypu-related, muskrat-resembling species of rodent found only in the West Indies. Several of these are notorious for having been written off as extinct, only to be unexpectedly rediscovered years later. Indeed, in two separate cases the species in

question was discovered alive several years after having been originally described from fossils.

Long ago, however, they shared their islands with some much bigger relatives, loosely termed giant hutias. According to traditional zoological dictum, all of these became extinct well before the West Indies were reached by Europeans, but there is some intriguing evidence to suggest otherwise – the apparent post-Columbus existence here of a curious creature known as the quemi.

This is the name of a mysterious rodent mentioned by explorer Gonzalo Fernández de Oviedo y Valdés in his 16th-Century account of Hispaniola. It was said to be brown in colour, like this island's hutias, but larger in size. Yet following Oviedo's report, nothing more was heard of the quemi – until the 1920s.

That was when some bones of a large, previously unknown species of rodent were discovered in a cave near a plantation at St Michel, Haiti, on Hispaniola. After studying them, Dr Gerrit Miller of the Smithsonian Institution identified their owner as a representative of Oviedo's obscure quemi, and in 1929, within his formal description of the bones, Miller named their species *Quemisia gravis.* Remains have since been found in the Dominican Republic on Hispaniola too. Moreover, this species was apparently a traditional item of food for the native Hispaniolans, as its limb bones have been found in early kitchen middens.

Today, the quemi is also known as the twisted-toothed giant hutia - but whatever happened to it? Researchers believe that this interesting rodent died out soon after the arrival on Hispaniola of the Spaniards, and certainly no later than the 16th Century's close – yet another irreplaceable island endemic that simply couldn't compete with the arrival of humankind and its ever-attendant array of introduced species, particularly the black rat.

Having said that, in 1989 one researcher speculated that Oviedo's description of the quemi may not have been a reference to *Quemisia gravis* after all, but instead to *Plagiodontia velozi* (aka *P. ipnaeum*). This is a now-extinct species of Hispaniolan hutia known as the Samana hutia, whose remains have been found with those of black rats, thus suggesting that it was still alive when the first Europeans and their stowaway rodent entourage first reached this Caribbean island in 1492.

THE CRYPTIC COMADREJA AND THE SPURIOUS SPANISH RACCOON
The Samana hutia may also (or alternatively) be the identity of a second mystifying, still-unclassified Hispaniolan rodent. Known locally as the comadreja, this cryptid allegedly survived here until the 20th Century.

Yet another Caribbean mystery mammal that may be hutia-related is the so-called 'Spanish racoon' mentioned in Dr Patrick Browne's *The Civil and Natural History of Jamaica* (1756). Including it in a listing of six species of mammal that he collectively termed Mus (which is the genus housing the common, typical species of mouse within the taxonomic family Muridae),

An engraving from 1894 depicting the Cuban hutia (aka the hutia-conga) *Capromys pilorides* – up to 3 ft long, it is the largest true hutia, but is much smaller than the now-extinct giant hutias

Browne claimed that this creature was not native to Jamaica but was frequently imported there from Cuba (where it was very common). He stated that it sported fairly rough fur; rabbit-like eyes, teeth, and lips, but wider nostrils, and shorter, smaller ears; plus a straight, tapering, hairy tail; and that it exhibited a vegetarian diet.

Interestingly, raccoons did formerly exist in Jamaica (and Cuba), but they were exterminated there by Spanish colonists who hunted them for their meat, with the last sightings reported in 1687 (and they were wiped out even earlier in Hispaniola, by 1513). However, no raccoon species corresponds with the verbal portrait by Browne given above, or is vegetarian, and it would be decidedly odd to categorise a bona fide raccoon as a mouse. Conversely, the Spanish raccoon as described by Browne was evidently a rodent. Consequently, when referring briefly to this enigmatic animal within his standard work *Extinct and Vanishing Mammals of the Western Hemisphere* (1942), American mammalogist Dr Glover M. Allen speculated that "it was probably the larger Cuban hutia, *Capromys pilorides*". Although this identification is certainly very plausible, it has never been formally confirmed, so the mystery of Browne's 'Spanish racoon' remains officially unresolved.

Typical marmots, genus *Marmota*

African marmot engraving #1

African marmot engraving #2

African marmot engraving #3 (French)

IDENTIFYING THE MYSTIFYING AFRICAN MARMOT

I have random luck and coincidence to thank for introducing me to a most intriguing rodent riddle - the hitherto long-forgotten mystery of the African marmot.

During the evening of 1 October 2013, I was browsing online in search of some suitable vintage wildlife engravings to use for illustrating an article that I had just written when I came upon a remarkable hand-coloured engraving of an even more remarkable animal – one that I had never even heard of before. The engraving, reproduced here (African marmot engraving #1), was entitled 'Marmota Africana' – 'African marmot'. This immediately piqued my curiosity, for one very basic reason.

Marmots are large ground squirrels belonging to the genus *Marmota*, and are native to Europe, Asia, and North America – but not Africa. So whatever has happened to its missing marmot, because no such species is recognised by science today? Having said that, however, and as can readily be seen from this engraving, it doesn't actually look like a marmot, or any other kind of ground squirrel. On the contrary, its distinctive combination of very pale fur, extremely long, projecting teeth, and very large, spatulate, mole-like feet suggested to me that it belonged to a totally different rodent family, one whose members were evidently very fossorial in behaviour.

In short, the creature that I felt the engraved mystery rodent most closely resembled was some form of blesmol or African mole-rat, thereby belonging to the taxonomic family Bathyergidae. Exclusively African in modern-day distribution, the blesmols inhabit elaborate burrows, are almost entirely subterranean in lifestyle, and currently number some 22 species. These include, as the family's most famous and highly-specialised member, the naked mole-rat *Heterocephalus glaber* – one of only two species of eusocial mammal. (The other one, incidentally, is also an African mole-rat – the less-familiar Damaraland blesmol *Fukomys damarensis*.)

Researching this curious engraving online, I discovered that it dated from 1795, but I was unable to locate its original published origin. However, I did find a couple of other images of its enigmatic subject, though both of these, again reproduced here, were monochrome. All three images clearly portrayed the same species, and indeed, they were most probably nothing more than slightly altered or enhanced copies of a single original illustration.

One (engraving #2) was fundamentally a mirror-image version of the other two, and was clearly derived from a French-language publication, because it was labelled 'Marmotte Africaine'. The third one (engraving #3) dated again from 1795, and I was able to trace it to C.P. Thunberg's tome *Travels in Europe, Africa, and Asia*. Further investigations revealed that this same source was claimed for the colour engraving (engraving #1) too, upon which the less-detailed monochrome version (engraving #2) had presumably been based.

C.P. Thunberg was Carl Peter Thunberg (1743-1828), a Swedish naturalist who conducted very significant studies of South Africa's flora and fauna during three years' sojourn in Cape Colony. When I discovered this, I realised that the so-called African marmot was very similar in appearance to the Cape dune mole-rat *Bathyergus suillus* – sharing the same very pale fur, large spatulate feet, and long protruding teeth, plus its tiny eyes, sturdy body, and short tail. Could this stocky species, the largest of the blesmols, be the identity of Thunberg's mystery 'marmot'?

Researching the French engraving soon gave me my answer. I learnt that a copy of it was present online in a Wikipedia Commons page, and sure enough, it was entitled 'File:Bathyergus suillus as Marmotte Africaine.jpg'. Information accompanying this engraving on that page revealed that it had appeared in a 1796 French edition of Thunberg's travelogue, in which he had applied the misnomer 'African marmot' to the Cape dune mole-rat.

The mystery of an alleged *Marmota* species native to the Dark Continent was duly solved, with the said 'marmot' unmasked as a mole-rat. In other words, it was a curious but fascinating misidentification that had been forgotten by mainstream zoology for over 200 years, until that fateful evening when I had fortuitously come upon an engraving of the long-lost 'African marmot'.

Taxiderm specimen of the Cape dune mole-rat (Mariomassone/Wikipedia)

Chapter 12:
REVISITING THE KINGDOM OF THE TERROR BIRDS

I will tell also of the huge bird which chased Challenger to the shelter of the rocks one day - a great running bird, far taller than an ostrich, with a vulture-like neck and cruel head which made it a walking death. As Challenger climbed to safety one dart of that savage curving beak shore off the heel of his boot as if it had been cut with a chisel. This time at least modern weapons prevailed and the great creature... - phororachus [sic] its name, according to our panting but exultant Professor - went down before Lord Roxton's rifle in a flurry of waving feathers and kicking limbs, with two remorseless yellow eyes glaring up from the midst of it.

Sir Arthur Conan Doyle - *The Lost World*

During the two decades that have elapsed since the original version of this chapter appeared in my book *Extraordinary Animals Worldwide* (1991), some very significant, dramatic discoveries have been made in relation to its awe-inspiring subjects. High time, therefore, that we paid a return visit to the monstrous kingdom of the terror birds.

Imagine a scene in which terror-stricken horses and rhinoceroses are fleeing for their lives, pursued by a gigantic flightless bird with huge talons and an enormous beak with which it will mercilessly tear its hapless victims apart. Surely the nightmare of a diseased mind, or a clip from some outrageously over-the-top science-fiction film? On the contrary - for such scenes as this were part of everyday life on the North American plains during the early Eocene epoch, approximately 50 million years ago.

By then, the great dinosaurian dynasty had been extinct for over 14 million years. Many of the ecological niches or lifestyles requiring vegetarian species - once filled by its vast assemblage of herbivorous forms, such as *Diplodocus* and *Triceratops* - were now occupied by plant-eating mammals. In contrast, those niches calling for large-sized land carnivores - formerly occupied by such ferocious flesh-eating dinosaurs as *Allosaurus* and *Tyrannosaurus* - had yet to attract mammalian representation, because these furry animals would not beget any sizeable meat-eaters on land for another 10 million years or so.

Instead, it was the bird lineage that sought supremacy in terrestrial carnivory during the Eocene, and in so doing provided the greatest threat to the mammals since their evolution in the shadow of the dinosaurs more than 150 million years earlier.

THE DREADED DIATRYMIDS

The birds initially responsible for this daring attempt to undermine mammalian success belonged to a now-extinct taxonomic group formerly believed to be related to the cranes but now reassigned to the waterfowl, and known as the diatrymids or gastornithids. The first species, which evolved during the Palaeocene epoch, approximately 64-56 million years ago, were ground-dwelling birds of modest dimensions that probably descended from primitive, superficially crane-like forms, but eventually switched to an overtly carnivorous lifestyle, transforming into giant, stout-legged, earthbound birds whose wings had become too small to sustain flight. Europe housed three known species – *Gastornis parisiensis* (late Palaeocene to early Eocene), *G. russeli* (late Palaeocene), and *G. sarasini* (early Eocene to mid-Eocene). There was also a single Chinese species, *Zhongyuanus xichuanensis*, dating from the early Eocene and formally described in 1980, but it is currently represented by just a single tibiotarsus bone (the fused tibia and proximal tarsus bone in birds), uncovered in China's Honan Province during 1976.

However, the diatrymids attained their revolutionary zenith with North America's early-to-mid-Eocene representative - the most famous and spectacular diatrymid of all. This was *Diatryma giganteus*, sole member of the genus *Diatryma*, which was temporarily renamed *D. steini* after fossil collector William Stein, who had found the first virtually complete skeleton of this awesome species during the American Museum of Natural History's Palaeontological Expedition of 1916, and had duly presented it to the museum.

(Speaking of names: nowadays, due to the laws of nomenclatural precedence, *Diatryma giganteus* should by rights be referred to as *Gastornis giganteus*, because the two genera are no longer deemed sufficiently different to warrant separate names, and as *Gastornis* was coined a few years before *Diatryma* it takes precedence. However, because the only *Gastornis* species that this present chapter is concerned with is the one that was formerly housed within the genus *Diatryma*, I am retaining *Diatryma* here for purposes of clarity.)

An inhabitant of what is now Wyoming and New Mexico, there is no doubt that this huge bird must have been a terrifying sight to behold. With a total height of 6-7 ft, it towered over its mammalian counterparts in much the same way as a modern-day ostrich would have done. Whereas the latter is built upon relatively lissom, gracile lines, however, with long slender neck and small head (comparable, in fact, to the European *Gastornis* contingent), North America's *Diatryma* was a much more robust form.

Its limbs were proportionately shorter but much sturdier, its body was very bulky, its neck was extremely thick but relatively short, and its head was disproportionately large and laterally flattened, with a skull that measured 1.5 ft in total length - equalling that of a modern-day horse. And unlike anything exhibited by an ostrich, its beak was a massively constructed,

The very formidable *Diatryma giganteus* (Hodari Nundu aka Justin Case)

ferocious-looking weapon, almost as deep as the skull itself and with its upper half terminating in a sharp point. Its four-toed feet were equally noticeable, as each toe bore a curved, talon-like claw. In fact, the only insignificant features of this bird's morphology were its wings, which were little more than rudimentary stumps.

With no mammalian competitors, the *Diatryma* contingent of diatrymids were the supreme ground-dwelling predators of the northern hemisphere in Eocene times, inhabiting grasslands and hunting the primitive herbivores that constituted some of today's most familiar plant-eating mammals. Those early forms were much smaller than their modern descendants. For example, the horse's Eocene antecedent, *Hyracotherium* (=*Eohippus*), was no larger than a fox terrier; and *Hyrachyus*, a proto-rhinoceros, was not much bigger either. Little wonder, then, that they lived in fear of the diatrymas, which are thought to have been at least adequate runners, capable of seizing and killing any of this epoch's largest mammals, and then tearing off chunks of flesh from the carcase with their huge beak and claws. But were they?

A pair of magnificent, life-sized *Diatryma* models on display in Reutlingen, Germany, in 2003 (Markus Bühler)

Diatryma depicted upon a postage stamp issued in 1990 by the Yemen Republic

Some researchers have challenged this traditional image of the diatrymas' lifestyle, suggesting instead that they may actually have been placid herbivores themselves, using their beaks merely for slicing fruit or scything through tussocks of grass, and employing their talons only in defence - thereby likening these mighty birds to oversized cassowaries. Moreover, in a very recent study by Dr Delphine Angst and co-workers, published in April 2014 and analysing the carbon isotope composition of apatite in European *Gastornis* bones, no evidence was found for the presence of meat in their diet.

Other researchers, however, continue to uphold a predatory role for diatrymas, not only basing their opinions upon anatomical considerations, but also acknowledging that after the carnivorous dinosaurs had died out, the evolution of some type of large-sized terrestrial flesh-eater was inevitable, in order to maintain ecological balance. Additionally, there is the compelling fact that diatryma-like birds have evolved not once but twice - and on the second occasion there is no doubt whatsoever that they were bona fide meat-eaters.

ENTER THE PHORUSRHACIDS
By the Eocene's close, 34 million years ago, all diatrymids everywhere were extinct - out-competed by the emergence of large mammalian carnivores called creodonts. With no descendants to perpetuate their line, it seemed reasonable to suppose that diatryma-like birds would never again exist on Earth - but then came the giant phorusrhacids or terror birds.

At the end of the Cretaceous period around 64 million years ago, South America broke away from North America, becoming a massive island continent, just like Australia today. It did not reconnect with North America, via the isthmus of Panama's formation, until approximately 4.5 million years ago, during the Pliocene epoch. So for almost 60 million years, its wildlife pursued its own, unique lines of evolution, predominantly independent of those on other continents, sealed off as it was from large-scale invasion by terrestrial mammals from elsewhere. During this immense timespan, South America's mammals engendered a vast array of large herbivores - but surprisingly few carnivores. In fact, its only meat-eaters of comparable stature were the borhyaenids - marsupials variously resembling bears, wolves, and even sabre-toothed cats (but nowadays usually housed in their own taxonomic order, Sparassodonta, separate from all true marsupials). And so, to restore ecological equilibrium, the phorusrhacids evolved.

More closely related to the cranes than the diatrymids were, the phorusrhacids were believed until quite recently to have first appeared during the late Oligocene epoch, around 26 million years ago. However, their earliest known representative is now *Paleopsilopterus itaboraiensis*, a Brazilian species dating from the mid-Palaeocene epoch, 60 million years ago, which was formally described in 1985. Just like the diatrymids, the phorusrhacids began as moderately proportioned ground-dwelling birds, whose wings gradually decreased in size as the birds' overall size increased - dramatically.

Before the extinction of their most recent known member, in Uruguay during the late Pleistocene epoch around 17,000 years ago, the phorusrhacids had become enormous, represented by monster-sized species standing around 9 ft tall (i.e. at least 1 ft taller than the modern-day ostrich). Probably the most famous phorusrhacid, however, was the medium-sized *Phorusrhacos* (aka *Phorusrhacus* and *Phororhacos*) *inflatus*, a mid-Miocene species that stood 5-6 ft tall, and inhabited Patagonia's plains 15 million years ago (which were later home to the 6.5-7-ft-tall *P. longissimus*).

The first phorusrhacid remains were unearthed in 1887 by Argentinian palaeontologist Prof. Florentino Ameghino. Unfortunately, these were not complete or near-complete skeletons, but only small, isolated fragments. Because of this, one portion - part of a phorusrhacid's lower jaw - was misidentified by Ameghino as a fragment from a prehistoric relative of the armadillos and anteaters. Happily, more extensive remains were found in 1891, which revealed that they were in truth from gigantic birds.

Very remarkably, since the phorusrhacids did not descend from them and were not even closely related to them, their overall appearance compared well with that of the diatrymids, especially the powerful diatrymas. All of the diatryma hallmarks were present - huge size,

South America's Pliocene procyonid *Chapalmalania* confronted by a phorusrhacid (Hodari Nundu aka Justin Case/Deviantart)

sturdy neck and legs, four-toed feet bearing formidable talons, extremely large head with enormous beak (which was sharply hooked in the phorusrhacids), and tiny useless wings. Having said that, the *Phorusrhacos* species were rather more slender than *Diatryma*, with slimmer, proportionately longer legs, indicating that they were faster, more efficient runners. Some closely allied species, however, notably the monstrous 'bird of thunder' *Brontornis* (see p. 124), were much more robust, thereby providing a closer diatryma duplication.

Yet whereas controversy exists concerning whether the diatrymas were predatory, there can be no doubt at all on that score with the phorusrhacids. In contrast to the parrot-like appearance of the diatrymas' huge beaks, those of the phorusrhacids were more like an eagle's, due to the extremely pronounced hook on their upper mandible, which was undoubtedly used to seize prey, as with modern-day birds of prey. Also their talons were larger and more powerful than those of the diatrymas, and were clearly designed for ripping flesh. Nevertheless, their overall comparability to diatrymas is sufficiently marked to imply strongly that the two groups fulfilled the same ecological role. So if, as seems certain, the phorusrhacids were flesh-eaters, the diatrymas must surely have been too (but perhaps less effective ones?).

Intriguingly, the phorusrhacids are one of the very few major groups of fossil birds to have attracted anything remotely approaching the level of public popularity enjoyed by such prehistoric stalwarts as the dinosaurs, mammoths, and sabre-toothed cats - a feat no doubt accomplished by virtue of their nightmarish appearance and terrifying lifestyle. Notably, one of the most impressive of the many breathtaking exhibits at the spectacular 'Dinosaurland' exhibition at Drayton Manor Park, in Worcestershire, England, is a full-scale, life-like model of a *Phorusrhacos*, vividly recapturing and restoring the chilling vision that confronted South America's herbivorous mammals in bygone days.

Similar *Phorusrhacos* models are also displayed at comparable exhibitions elsewhere, including the 'Valley of the Dinosaurs' at Wookey Hole in Somerset, England.

A phorusrhacid even stars in at least one major science-fiction film, *Mysterious Island* (Columbia/Ameran films, 1961), based very loosely upon the eponymous Jules Verne novel. That master of movie monsters Ray Harryhausen created a very impressive animated model of a *Phorusrhacos*, complete with small wings and huge hooked beak, which duly did its best to decimate the film's heroes before being subdued and transformed humiliatingly into their next meal. Tragically, however, the novelty value of using a *Phorusrhacos* adversary (rather than the customary dinosaur or sea monster) was lost on the film audiences of that time. This was because its true identity was never revealed by any of the film's characters - so the audience simply thought that it was a giant chicken!

In 2006, James Robert Smith's novel *The Flock* was published, a contemporary eco-thriller in which a remote swamp in Florida is home to a secretive, surviving flock of phorusrhacids, which find their domain threatened by corporate plans to develop it as the site for a theme park. Needless to say, the phorusrhacids are not best pleased with this, and the foolhardiness of their human antagonists in attempting to pit themselves against a colony of super-intelligent modern-day terror birds soon becomes all too apparent. Toss in a billionaire rogue

The magnificent, life-sized *Phorusrhacos* model at Drayton Manor Park's 'Dinosaurland' exhibition (Dr Karl Shuker)

environmentalist committed to protecting them once their existence is discovered, plus an obsessive hunter with very different ideas, and a highly engrossing plot duly ensues – as I can personally testify, having purchased and read this novel soon after it was first published.

More recently, *The Flock* has been optioned for production as a movie, so the prospect of CGI terror birds stalking the big screen may soon become a reality – or '*Jurassic Park* meets *The Birds*', as some film buffs are already describing it. I can't wait!

Speaking of CGI terror birds, but this time on the small screen: In 2009, the third series of the popular British sci-fi television series *Primeval* included a thrilling episode that featured a rampaging flock of phorusrhacids on the loose in a Ministry of Defence zone after emerging through a time portal known as an anomaly linking their prehistoric realm to the present day.

THE TERROR OF *TITANIS*

The phorusrhacids were a longer-lasting group than the diatrymids, surviving in South America from

Holding my copy of *The Flock* while a very different bird of terror, the ferocious avian deity Garuda, looks on approvingly - I hope! (Dr Karl Shuker)

the Palaeocene right through to the Pliocene. So when the isthmus of Panama rose above sea-level during the Pliocene, rejoining South America to North America approximately 4.5 million years ago, they were given the opportunity to invade North America - a continent that had not seen anything like them since the extinction of its own diatrymas around 30 million years earlier. Moreover, palaeontologists have discovered that the phorusrhacid lineage did indeed reach North America, because during the early 1960s fossil bird expert Dr Pierce Brodkorb described a very sizeable new species of phorusrhacid based upon a tarsometatarsus (a bird's fused ankle/foot) and an associated toe bone that had been obtained by fossil collector Benjamin I. Waller from a site in Florida, USA (*Auk*, April 1963). And in 1995, a fossil toe bone directly comparable to the example from Florida was found in Texas and documented by palaeontologist Dr Jon Baskin.

Named *Titanis walleri*, this North American phorusrhacid was initially thought to date from the late Pleistocene (i.e. less than a million years ago), because its Florida remains had been found in association with the fossils of other species that definitely dated from that time period. So too had been the Texas toe. Moreover, *Titanis* had a 2-ft-long skull, and was originally believed to have stood 10-12 ft tall, but it has been downsized in more recent scientific estimates to a rather less dramatic though still very daunting 5-6 ft.

Back when it was thought to be a titan in every sense, however, phorusrhacid expert Dr Larry Marshall unsurprisingly dubbed *Titanis* 'the terror bird' (from which the entire phorusrhacid group duly derived their common, informal name). In a newspaper interview (*Columbus Dispatch*, Ohio, 12 February 1989), he was also quoted as considering it to have been "probably the most dangerous bird ever to have existed" (but that was before the truly gargantuan *Kelenken* was uncovered, as will be revealed here shortly!).

Following the invasion of South America by North American mammalian carnivores after the Panamanian isthmus had risen up, the phorusrhacids eventually died out there. As noted earlier, the most recent South American phorusrhacid currently known, dating from the late Pleistocene, is a species unearthed in Uruguay and documented in 1999. But what about *Titanis* – how long did it linger in North America?

Prior to the discovery of the Texas toe, the Florida remains of *Titanis* had already been re-dated as being around 2.5 million years old, but the Texas toe genuinely seemed to be of late Pleistocene age. Could it be, therefore, that this most monstrous of North American birds had actually been encountered by that continent's earliest human arrivals? If so, what a terrifying vision it would have been! In reality, however, further dating work, but this time applied directly to the toe (rather than using indirect methods, i.e. applied instead to the fossils associated with it, as had previously been done), revealed that it too, just like the Florida remains, was approximately 2.5 million years old. This in turn meant that *Titanis* and *Homo sapiens* had never met one another in North America after all. In addition, other *Titanis* remains were subsequently found here that dated back even further, being up to 5 million years old.

Yet as this meant that *Titanis* had been present there well *before* the Panamanian isthmus linking North America to South America had even arisen, how could such a very large flightless bird have reached the North from the South at that time? The only plausible method was by arriving on

A phorusrhacid depicted upon an unofficial Fujeira philatelic Miniature Sheet, issued in 2000

floating debris or vegetation – so *Titanis* hadn't utilised a land-bridge after all, but had simply floated there instead. Sadly for its lineage, however, even the mighty *Titanis* proved to be an inferior competitor to the plethora of large, ferocious mammalian carnivores present in North America, and it is believed that the youngest remains currently recorded for it are probably from some of the very last surviving individuals of its species.

Today, therefore, there are no phorusrhacids, either in the Americas or anywhere else - this awesome example of avian evolution is no more. However, they are not totally without modern-day representation. Flourishing alongside them throughout their history in South America was a closely-related sister group, the seriemas, which still have two species surviving here today - the 3-ft-tall crested seriema *Cariama cristata*, and the 2-ft-tall Burmeister's seriema *Chunga burmeisteri*. Although also allied to the cranes and bustards, they have slightly hooked beaks that recall those very much deadlier versions borne by the extinct phorusrhacids. In overall appearance, the seriemas also call to mind that long-limbed African serpent-slaughterer the secretary bird *Sagittarius serpentarius*, and seem to have a comparable lifestyle, preying upon snakes, lizards, small rodents, and a wide range of insects.

There is no doubt at all that although the phorusrhacids have been extinct for many millennia, there is still much to be discovered regarding these immensely impressive, imperious creatures - as proven during the mid-1980s, for instance, by a quite astounding revelation.

ANTARCTICA'S TERROR BIRD

During a scientific expedition spanning December 1986 and January 1987, California University palaeontologist Dr Michael O. Woodburne and his team uncovered a 3-in-long pre-maxilla section of a gigantic fossilised beak and a 3-in-thick fossilised tarsometatarsus. Both were from a phorusrhacid, but they had not been found in South America, and not in North America either. In fact, they had been discovered on Seymour Island - just off the north-east coast of the Peninsula of Antarctica!

19th-Century engraving of a crested seriema

Modern reconstruction of South
America's *Phorusrhacos* (Tim Morris)

Subsequently documented by Woodburne and colleagues in various *Antarctic Journal* reports, this new species was the first phorusrhacid known from outside the Americas. Estimated to have stood 6-6.5 ft tall (and probably a predator of the marsupial species found in the same rock strata), the Antarctic phorusrhacid attracted great scientific interest, because its very existence effectively disrupted all previously held ideas regarding phorusrhacid evolution. After all, how could science explain the presence of a huge flightless bird (also unable to swim) not simply in the Americas but also in Antarctica - a totally isolated world surrounded on all sides by the vast oceans?

Contrary to popular assumption, Antarctica has not always been the ice-encapsulated continent that it is today. Until it drifted towards the South Pole a few million years ago, it was a thriving world teeming with wildlife and bearing vast expanses of lush tropical forest, just like South America. This has been revealed by the ever-increasing number and variety of fossil species of animal and plant discovered here (especially on Seymour Island) since the onset of the 20th Century - but that is not all.

A terror bird taking on a sabre-tooth duo (Hodari Nundu aka Justin Case/ Deviantart)

Researchers interested in the relative positions of Earth's continents millions of years ago have traditionally believed that Antarctica became an island continent at much the same time as South America did, about 64 million years ago. This would mean that all terrestrial, flightless animals (plus those incapable of strong, sustained swimming or rafting on floating mini-islands) evolving in Antarctica after that time would be unique, occurring nowhere else in the world. This is because they would be physically unable to escape from their insular Antarctic homeland - thus paralleling the situation with animals in South America during its own period of island existence, as described earlier.

However, in later times this belief has been seriously challenged, because some fossil species have been found in Antarctica that are very closely related to others living in South America as little as 40 million years ago, during the Oligocene's onset. This means that the two continents must have been joined (probably by an interconnecting land-bridge like the isthmus of Panama) for at least 24 million years longer than previously supposed.

Moreover, it is now believed that certain animals that were once thought to have first arisen in South America actually originated in Antarctica, only later migrating into South America via this land-bridge. This belief has arisen because the fossils of these animals found in Antarctica have proved to be older and more primitive than those from South America.

Those ideas duly gained substantial extra support with the Antarctic phorusrhacid's discovery, which is why it was so important a find. Until then, the oldest known phorusrhacid fossils were all from South America, and only dated back to the Oligocene, 35 million years ago. However, the La Meseta formation of rock on Seymour Island in which the Antarctic specimen's remains were discovered dates back a further 5 million years, to the late Eocene. Moreover, some fossilised bird footprints found in the early 1980s on Antarctica's King George Island and now thought to have been made by phorusrhacids are at least 5 million years older than the beak and ankle bone.

In short, based upon fossil evidence on record at that time, phorusrhacids existed in Antarctica at least 10 million years before their first known occurrence in South America. Thus, unless there were even older fossils still awaiting discovery in South America, it seemed that the phorusrhacids did not originate there, as hitherto believed, but in Antarctica instead, reaching South America later via a land-bridge. However, as already pointed out, South American phorusrhacid fossils dating from the Palaeocene are now known, and thus pre-date the Antarctic specimen, so this theory is no longer tenable based upon physical evidence currently on record.

Nevertheless, just by being there, Antarctica's very own terror bird (which has still to be formally named and described, incidentally) has unfurled a previously unsuspected and extremely significant episode in the phorusrhacids' evolution.

TRANSATLANTIC TERROR BIRDS?
Nor was the Antarctic phorusrhacid the only zoogeographical surprise that the terror birds have had in store for palaeontologists lately. As recently as 27 November 2013, a team of French researchers featuring veteran palaeontologist Dr Eric Buffetaut published a paper in the online scientific journal *PLOS ONE* announcing that their re-examination of some French (Lissieu) and Swiss (Egerkingen) avian fossils previously categorised as diatrymids and ratites respectively, and dating from the mid-Eocene, were in fact from a large phorusrhacid, to which the team has ascribed the name *Eleutherornis cotei*.

Suddenly, the phorusrhacids were indigenous to the Old World too, not just to the New World and Antarctica. True, there had been attempts previously to identify as phorusrhacids the Paleogene-dated remains of certain smaller European fossil birds known as ameghinornithids, but these were subsequently reclassified as basal seriemas instead. Yet how had phorusrhacids managed to reach Europe, and at such an early geological date?

The answer may rest with the fossilised remains of another ostensibly out-of-place terror bird – *Lavocatavis africana*. For as its name reveals, this mysterious species was uncovered in Africa. In 2011, celebrated fossil bird expert Dr Cécile Mourer-Chauviré headed a team of researchers who published a paper in the journal *Naturwissenschaften* that formally described and named *Lavocatavis*. It was represented by only a single femur, but one that exhibited a specific combination of features that collectively indicated a phorusrhacid identity for its long-demised owner, which had lived during the early or early-to-mid-Eocene (i.e. between 52 million years and 46 million years ago) in what is today southwestern Algeria. But how could

Did flightless South American phorusrhacids island-hop across the Atlantic to Africa?

the phorusrhacid lineage have reached Africa – far away indeed from the land masses that became today's New World and Antarctica? The Mourer-Chauviré team proposed two alternative hypotheses to account for this.

One of these speculated that as phorusrhacids had originated from flying birds, perhaps a flying ancestor had reached Africa from the New World and then, via convergent evolution, had given rise to larger flightless forms whose morphology closely paralleled their New World equivalents. The other, more plausible hypothesis was that a transoceanic migration of flightless phorusrhacids had occurred from South America to Africa during the Paleocene or very early Eocene. As the team pointed out, paleogeographic reconstructions of the South Atlantic Ocean suggest the existence of several islands of considerable size between South America and Africa during the early Tertiary Period, which could have assisted a transatlantic dispersal of such birds.

But what about Europe's *Eleutherornis cotei*? Buffetaut and his team suggested that if *Lavocatavis* were truly a terror bird, then Eocene-dated African phorusrhacids could have reached Europe by dispersal across the Tethys Sea. Several other groups of land vertebrate are known to have achieved this during the same time period, thanks to the presence of certain 'stepping stone' areas of intervening land, such as the Alboran and Apulian platforms, particularly during periods of low sea level.

Naturally, all of this involves a great deal of speculation, but the European and African fossil evidence noted above is real and appears to have been reliably identified as being of phorusrhacid origin. So it would seem that, somehow, the terror birds did indeed become transatlantic in zoogeographical distribution during their early evolutionary development, but we must await further palaeontological finds before we can obtain a clearer picture of how they achieved this remarkable feat.

A HANDS-ON WINGS THEORY FOR THE TERROR BIRDS – LITERALLY!
During the early 1990s, palaeontologist Dr Robert Chandler began seeking *Titanis* fossils in the Santa Fe River passing through Florida. During one search, he found a carpometacarpus bone (the avian equivalent of a hand), and part of an unexpectedly strong, robust-looking humerus (upper arm bone). This discovery was very important, because phorusrhacid remains had been notable until then for their absence of wing skeleture, leading researchers to assume that during the course of evolution these birds' wings had degenerated until they had eventually vanished altogether. Chandler's find, however, proved that this was not so, as the wing was hardly vestigial, measuring 3 ft long.

But it was its structure that so amazed him, because the fused bones constituting the carpometacarpus possessed what looked like a large round attachment region for a very sizeable flexible thumb. Adding to this the noticeably thick build of the humerus, in 1994 Chandler speculated that in *Titanis* each of its two wings may have actually evolved into an arm, equipped with a non-folding, permanently outstretched clawed 'hand' or 'paw' - which in turn may have also possessed at least a second (and possibly even a third) smaller, fixed clawed digit in addition to its much larger, flexible clawed thumb. Chandler also suggested that *Titanis* may therefore have used its 'arms' to prevent its prey victims from attacking it with their horns or hoofs, and also that it could manipulate its victims with its 'hands', and impale them with its claws. It would thus behave in a manner unexpectedly reminiscent of certain bipedal carnivorous dinosaurs, such as the infamous *Velociraptor*.

This fascinating theory engendered all manner of inspired visual representations of claw-

Titanis depicted with clawed 'hand-arm' wings (Tim Morris)

armed phorusrhacids during the next few years. In addition, during the late 1990s Czech cryptozoologist Dr Jaroslav Mareš boldly proposed in his book _Svet Tajemných Zvírat_ (1997) that a living, late-existing representative of _Titanis_ may even have inspired various traditional Native American legends of Raven. This was a giant mythical bird with a massive hooked beak, long sturdy legs on which it ran rapidly across the ground, and a feathered body like other birds - but possessing front paws with claws, instead of wings.

Sadly, however, clawed, pawed phorusrhacids proved to be a concept of short tenure. When the wing structure of the phorusrhacids' closest living relatives, the seriemas, was examined in a detailed review published during 2005, the researchers discovered that it too possessed a large round attachment region – but no claws. Clearly, then, there was no imperative after all that this region on the phorusrachids' wings must have been clawed either. And even if they had been clawed and thus resembled hands, there was no tangible evidence that _Titanis_ held its wings outstretched - or that they were proportionately robust. On the contrary, its wings turned out to be proportionately small. And so it was that another once-promising, highly-ingenious theory swiftly and silently disappeared from the scientific literature.

KELENKEN AND _BRONTORNIS_ – THE BIGGEST TERROR BIRDS OF ALL TIME?
Titanis may have been dethroned as the world's mightiest terror bird by recent downward revisions of its likely size, but a worthy new successor would soon be making its debut on the

Height comparison of *Homo sapiens* alongside (from left to right) *Kelenken guiller-moi, Phorusrhacos longissimus, Titanis walleri,* and *Gastornis parisiensis*

scientific stage. In 2007, a hitherto-unknown but truly spectacular species (and genus) of phorusrhacid from Argentina's Patagonia region was formally described and named – *Kelenken guillermoi*. Living some 15 million years ago during the mid-Miocene, with its genus named after an Amerindian winged deity, *Kelenken* is estimated to have stood 9-10 ft tall (its tarsometarsus

A delightful (if deluded!) illustration of *Brontornis* doing battle with an unidentifi-able reptile, c.1902-1906

bone alone measured around 18 in long). Furthermore, its 28-in-long skull (which included an 18-in hooked beak) is the largest ever recorded from any known species of bird, past or present.

Also greatly deserving of mention here is Burmeister's 'terror bird of thunder' – *Brontornis burmeisteri*. Another Patagonian phorusrhacid of the Miocene epoch, but probably preceding *Kelenken*, and known from an assemblage of limb bones, skull bones, and vertebrae, *Brontornis* is presently thought to have stood around 9.2 ft tall, thus making it only a little shorter than *Kelenken*, but it was much bulkier and heavier. Indeed, with an estimated weight of 770-880 lb, *Brontornis* is popularly believed to have been the third heaviest known bird of all time, exceeded only by Madagascar's modern-day great elephant bird *Aepyornis maximus* and by Australia's giant mihirung or 'demon duck of doom' *Dromornis stirtoni* from the late Miocene to the early Pliocene.

And speaking of demon ducks of doom: in 2007, an intriguing new study of its anatomy proposed that *Brontornis* was not a phorusrhacid at all, but was instead a monstrous, basal form of anseriform, i.e. a member of the taxonomic order housing the waterfowl (ducks, geese, and swans), as are the mihirungs. So its status as a terror bird may soon be over. Whatever its taxonomy, however, due to its huge weight *Brontornis* probably exhibited a hunting mode somewhat midway between active pursuit of its prey and ambush predation.

As for *Kelenken*: whether it actively pursued and killed its own prey, or simply chased off other carnivores from theirs instead and then devoured that, is presently uncertain. What *is* certain, however, is that this most formidable of phorusrhacids would have been a truly terrifying sight to behold, especially for any smaller, less powerful mid-Miocene mammals that had the misfortune to inhabit its Patagonian domain – a veritable personification, in fact, of the epithet 'terror bird'.

Summing up, it appears that the mammals' ultimate supremacy was indeed hard won; that with foes as devastating as the diatrymids and phorusrhacids, the threat to their survival was every bit as daunting as the one that they had already faced from the dinosaurs. Nevertheless, the dominant position held today by the mammals (which of course include ourselves) reveals unambiguously that they did win after all, and that the birds' bold, audacious bids to overthrow the Age of Mammals failed - but only just?

EPILOGUE: THE HOBBITS OF FLORES VS THE GIANT STORK OF DOOM?
Although no new dynasty from the avian lineage has arisen to restore the widespread supremacy formerly enjoyed by the phorushacids, a few much more limited examples may have occurred in quite recent times. One possible contender is what I have dubbed the giant stork of doom!

Where on Earth would you find all within the very selfsame location a unique, Alice-in-Wonderland-reminiscent assemblage of mega-rats and mini-humans, gargantuan lizards, dwarf elephants, and gigantic storks? Only on Flores – a small island in Indonesia's Lesser Sundas group that provides some classic examples of both island gigantism and insular dwarfism.

Between 20,000 and 10,000 years ago, it was home to: two extra-large species of giant rat, *Papagomys armandvillei* and *P. theodorverhoeveni*, the former of which still survives today (and indeed, according to some researchers the latter may do too); a dwarf subspecies of elephant-related

Homo floresiensis vs *Leptoptilos robustus* – were the hobbits of Flores stalked by a stork? (Hodari Nandu aka Justin Case/Deviantart)

stegodont proboscidean *Stegodon florensis insularis*, which died out around 12,000 years ago and is the youngest stegodont form on record from southeast Asia; the Komodo dragon *Varanus komodoensis*, the world's largest living species of lizard, which still survives on Flores today (as well as on the neighbouring islands of Komodo, Rintja, and Padar) and undoubtedly preyed upon Flores's dwarf stegodonts; plus the two highly significant species featured in this epilogue.

One of these species is the so-called 'hobbit' – the informal nickname given to Flores Man *Homo floresiensis*. With its most complete recorded specimen estimated at around 18,000 years old, this controversial, diminutive species of hominid apparently stood little more than 3 ft tall, may be descended from *Homo erectus*, and is believed to represent another case of insular dwarfism. Its first scientifically-documented remains were discovered in September 2004 at Liang Bua Cave in western Flores.

And the other species, whose remains were also uncovered in that very same cave, may have been one of the Flores hobbits' deadliest antagonists. It is a marabou stork, but far bigger than any of today's trio of marabou species. Yet it remained undescribed by science until as recently as 2010. Formally dubbed *Leptoptilos robustus*, the giant stork of Flores stood approximately 6 ft tall and weighed up to 35 lb, due to its extremely heavy leg bones – on account of which it is likely to have been predominantly if not exclusively flightless.

Even so, Smithsonian Institution palaeontologist Dr Hanneke J.M. Meijer and Dr Rokus A. Due from Jakarta's National Center for Archaeology, who jointly described its fossilised remains (dated

If it can't see me, it can't eat me, right? Wrong! Don't under-estimate its highly-developed sense of smell! Posing with a life-sized Komodo dragon model at Chester Zoo, England (Dr Karl Shuker)

at between 20,000 and 50,000 years old) in a *Zoological Journal of the Linnean Society* paper, believe that it was still theoretically capable of hunting and devouring juvenile hobbits. Having said that, there is presently no direct evidence confirming such activity; in the words of Dr Meijer as spoken during a 2010 interview with the UK's BBC television service: "Whether or not this animal may have eaten hobbits is speculative: there is no evidence for that." Then again, as Dr Meijer also conceded in that same interview: "But can not be excluded either."

In European tradition, the stork brings newborn human babies to their parents - but in a major reversal of that benevolent role, Indonesia's giant stork of doom may conceivably have hunted down such babies (possibly even older infants too) and devoured them. Only on the topsy-turvy island of Flores!

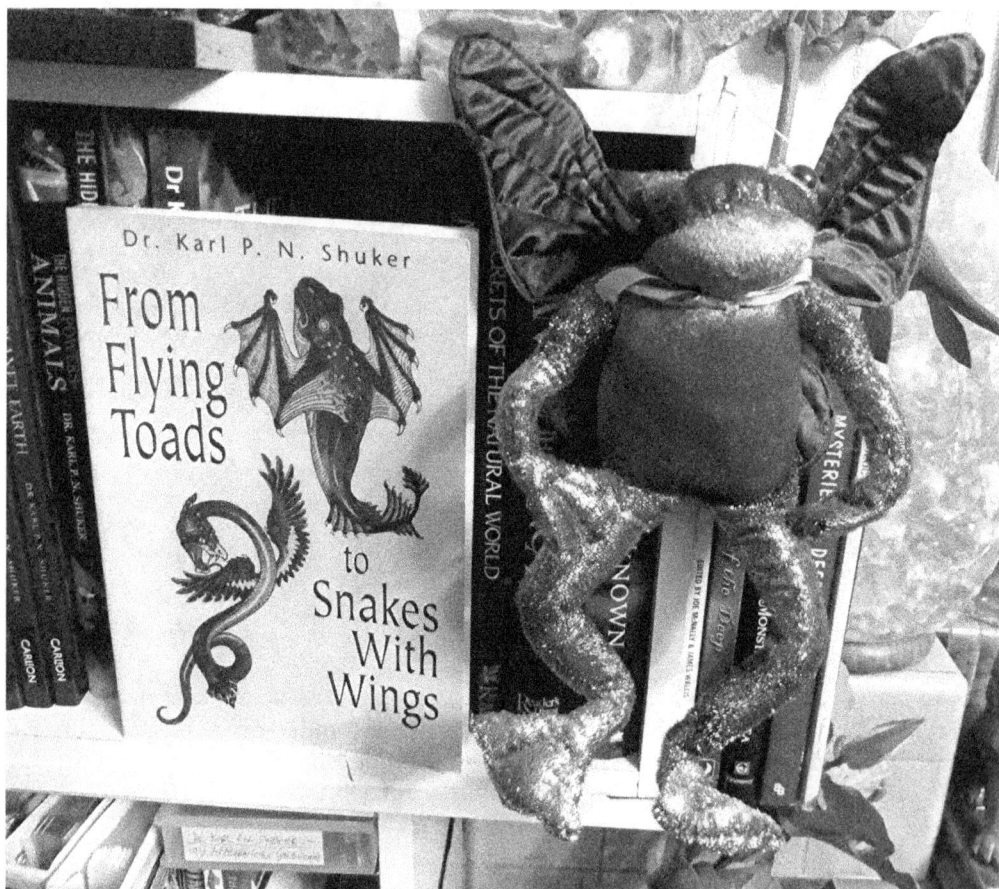

My book *From Flying Toads To Snakes With Wings*, featuring a representation of the Welsh water-leaper on its front cover, alongside a flying toad plush toy (Dr Karl Shuker)

Chapter 13:
TOADS WITH WINGS
- AND OTHER ODD THINGS

One of the most curious and interesting reptiles [in Wallace's time, the 1800s, amphibians were still categorised as reptiles] which I met with in Borneo was a large tree-frog, which was brought me by one of the Chinese workmen. He assured me that he had seen it come down, in a slanting direction, from a high tree, as if it flew. On examining it, I found the toes very long and fully webbed to their very extremity, so that when expanded they offered a surface much larger than that of the body. The forelegs were also bordered by a membrane, and the body was capable of considerable inflation. The back and limbs were of a very deep shining green colour, the under surface and the inner toes yellow, while the webs were black, rayed with yellow. The body was about four inches long, while the webs of each hind foot, when fully expanded, covered a surface of four square inches, and the webs of all the feet together about twelve square inches. As the extremities of the toes have dilated discs for adhesion, showing the creature to be a true tree-frog, it is difficult to imagine that this immense membrane of the toes can be for the purpose of swimming only, and the account of the Chinaman, that it flew down from the tree, becomes more credible. This is, I believe, the first instance known of a 'flying frog' [and its species was subsequently christened *Rhacophorus nigropalmatus*].

Alfred Russel Wallace – *The Malay Archipelago*

At the time of my writing it, as far as I was aware the flying toad referred to in the title of my book *From Flying Toads To Snakes With Wings* (1997) was an obscure creature of vaguely bufonine appearance and airborne ability that featured briefly in Welsh folklore and was known as the water-leaper or llamhigyn y dwr. In more recent times, however, two noteworthy developments have taken place that have added appreciably to its case file. One is the first attempt that I know of to identify this bizarre entity as a real animal, whereas the other is a revelation that strongly suggests it is little (if anything) more than a modern-day yarn, with no firm basis in either cryptozoology or traditional Welsh mythology. Additionally, I have recently learnt of an extraordinary report from France regarding some creatures that were described as bona fide flying toads, not merely cryptids that vaguely resembled such animals, plus a comparably cryptozoological record from England. Consequently, it is clearly time to revisit these curious aerial anurans.

A RAY OF HOPE FOR THE WELSH WATER-LEAPER?

Here is what I wrote about Wales's decidedly grotesque water-leaper in my *Flying Toads* book:

> One of the most formidable water monsters documented in John Rhys's exhaustive two-volume opus *Celtic Folk-Lore, Welsh and Manx* (1901) was the terrifying *llamhigyn y dwr* or water-leaper, which inhabited lonely stretches of river in Wales and devoured any hapless sheep or other livestock venturing into its freshwater domain. Its body's shape recalled that of a huge toad, but there the resemblance ended. Despite its name, the water-leaper lacked the toad's muscular, hopping hind legs. Instead, it had a tail - and a large pair of wings!
>
> Nevertheless, this weird wonder did share one notable characteristic with the toad - a powerful voice, literally a hideous shriek, with which it deliberately frightened so thoroughly any unwary travellers seeking to cross the river that they would lose their footing and fall headlong into it. Once in the river, they would swiftly make a brief (and invariably fatal) acquaintance with its monstrous occupant. Happily, the loathsome *llamhigyn y dwr* does not appear to have been encountered lately.

Certain subsequent accounts claimed that the water-leaper's tail bore a sting at its tip, but otherwise the above couple of paragraphs contained every detail of note regarding this thoroughly extraordinary beast because, very curiously, any stories or legends specifically featuring it were conspicuous only by their absence.

Not surprisingly, therefore, no attempt had ever been made to identify it as a real-life creature – which is why, on 22 November 2009, I was so intrigued by a short post on the CFZ bloggo that had been written by American cryptozoological enthusiast Dale Drinnon.

In it, Dale had offered a highly original, if equally unexpected, identity for the water-leaper – namely, a species of freshwater stingray. He sought to justify his claim by stating that stingrays have faces like toads, with raised eyes, and that although they lack legs they do leap out of the water sometimes, as a means of ridding themselves of parasites.

For me, the idea that the water-leaper could actually be a scientifically-undescribed species of freshwater stingray indigenous to the rivers of Wales is one of those remarkable notions that you just want so much to be true. How wonderful it would be if such a creature did exist – and in fairness there is a superficial similarity between the water-leaper's description and that of a stingray. Even so, for a cryptid identity to be plausible, it requires more than just a morphological match – it must also be compatible with the sociology and history of the cryptid's provenance. And such compatibility is sadly lacking in this instance.

In short: if anything as zoologically unexpected and as morphologically distinctive as a freshwater stingray – especially one that made its presence even more noticeable by leaping up out of its river habitat – had ever existed in Wales, the scientific world would have definitely known all about it long ago. In terms of natural history, the island of Great Britain, of which

Artistic representation of the Welsh water-leaper (Andy Paciorek / http://www.batcow.co.uk/strangelands/)

the principality of Wales is just one part, is among the most intensively-studied localities on Earth. Scarcely a blade of grass or square inch of soil has not been examined at some stage by generations of amateur naturalists, zoologists, and field biologists; ditto for its lakes, rivers, streams, shores, and coastlines.

Added to this is the preponderance of Britain's so-called "huntin', shootin' and fishin'" fraternity, which in its time has indeed hunted, shot, and fished anything of even modest size here, and preserved the remains afterwards as trophies. If a species of freshwater stingray had existed even in just a single Welsh river, every angler from miles around would have known of it and would have attempted to catch one. Equally, it would have been extensively documented in fishing journals and other natural history publications, there would have been display cases containing taxiderm specimens in country houses, pickled examples in museums, and probably even a fair scattering of local place-names commemorating such an exotic, eyecatching member of the local fauna.

A spotted species of stingray (Dr Karl Shuker)

True, it may well have been fished into extinction long ago, but records of such a nature as those just listed here would have survived to confirm its erstwhile existence, just as there are of other former fauna here, from bears and beavers to wolves, wild boars (before their recent reintroduction via escapes from captivity), and even a few possible references to the lynx. Instead, there is nothing whatsoever of this kind.

In any case, we also have to consider the zoogeographical enigma posed by the concept of a species of freshwater stingray living in Wales when no such fish or even any close relative of it is present anywhere else in Europe. Short of it dropping from the sky, how could such a specialised fish's existence in Wales but nowhere else for thousands of miles in any direction be explained by science? To quote Wikipedia's page on river stingrays:

> They are native to northern, central and eastern South America, living in rivers that drain into the Caribbean, and into the Atlantic as far south as the Río de la Plata in Argentina. Generally, each species is native to a single river basin, and the greatest species diversity can be found in the Amazon.

In addition, freshwater stingrays of the genus *Himantura* occur in various southern and southeast Asian rivers, including the Ganges, Mekong, and Chao Phraya.

Thus the possible existence of an unknown species of freshwater stingray in South America or Asia is not an unreasonable prospect. I might also look benevolently upon a distinctive new species inhabiting some remote river in New Guinea or even tropical Africa – but not in Wales or indeed anywhere else in the exceptionally well-traversed, much-scrutinised island of Great Britain.

And so, albeit with genuine sadness that such an awesome creature as a native British freshwater stingray just could never be, I must reluctantly conclude that there is no such blighter in good old Blighty!

Ironically, any detailed assessment of the Welsh water-leaper's possible zoological identity may be wholly surplus to requirements anyway. This is because there is now good evidence to suggest that far from being a valid (albeit exceedingly minor) creature of traditional Welsh folklore, the fearsome llamhigyn y dwr is of much more modern, and far more dubious, origin.

Once again, this revelation appeared in a CFZ bloggo post, this time by CFZ researcher Oll Lewis, who has a longstanding interest in Welsh lore and legend. On 6 March 2009, Oll firstly divulged how the famous story of the martyred dog Gellert - who allegedly rescued from the jaws of a wolf the baby son of Prince Llewellyn the Great of Wales, only for the prince to slay the poor hound in the mistaken belief that it had attacked his son – was a complete fabrication, a memorable yarn deftly spun by the owners of the inn in the North Wales town of Bedd Gellert to entice tourists to visit it. And instead of the town having derived its name from the dog Gellert, as often claimed in Welsh lore, it is now thought to have been named after Kellert, an 8th-Century Christian missionary, the site of whose grave is here.

Oll then revealed that the only known eyewitness account of the water-leaper ever recorded in print just so happened to have originated from a descendant of the same family who had invented the Gellert story, the descendant being a man named William Jones. He was the direct source of the water-leaper eyewitness account documented by John Rhys in his book of Celtic folklore that I referred to at the beginning of this present chapter's coverage of the water-leaper. The alleged encounter had taken place sometime during the first few decades of the 19th Century (Oll mistakenly stated the 18th Century), and featured a local eyewitness called Han Owen.

This is Oll's summary of Owen's encounter with the dreaded water-leaper:

> A local fisherman called Ifan Owen, also known as Han, had had an awful day's fishing.
>
> Whenever Han had cast out that day something had nibbled at his bait and removed it cleanly from the hook without getting snagged on it or pulling at the line. Because Han made his living from fishing he grew steadily more annoyed each time his bait was stolen and eventually, when he could take it no more he moved to another spot, beside a small cliff in the valley.

When Han cast his line out here he felt something pull at his bait almost instantly and, not wanting to lose any more of his bait, he pulled his rod back much more sharply than usual to be sure of hooking the animal that had been getting away with his bait all day.

Having finally hooked his tormentor Han had to pull with all his might to get the creature out of the water. After an epic struggle the monster erupted out of the water [and] it shot off the hook towards the cliff, so fast that, according to Han:

"It dashed so against the cliff that it blazed like lightning".

Han later recounted that if it were not the Llamhigyn then it must have been the devil himself.

Han claimed that both he and his father before him had seen the water leaper on several occasions and in a number of places along this stretch of water and it was said to scream loudly whenever a fisherman was able to pull it to the surface.

It all sounds highly implausible, but this is hardly surprising, because as Oll went on to disclose, the family who owned the Bedd Gellert inn:

...held regular tall-storytelling nights with relatives and friends from Bedd Gellert and the nearby parish of Dolwyddelen. Han Owen was regularly in demand for these nights as he had few equals in the area in his ability to spin a yarn and it was at one of these nights that William Jones first heard Han Owen delivering the tales of his encounters with the water leaper.

Given the dubious pedigree of the tale the smart money is on the water leaper having been one of Han's tall tales, but you can make up your own mind about that.

After reading this, I already have, and it's not good news for the veracity of the water-leaper as a genuine beast of traditional Welsh folklore – especially as every source of information concerning it that I have consulted can ultimately be traced back to Rhys's Celtic folklore book...whose own source was William Jones and, via him, the clearly unreliable eyewitness Han Owen.

Consequently, I would be extremely interested to learn if anyone knows of any publication documenting the water-leaper whose publication precedes that of Rhys's book and whose source is not the Han Owen eyewitness account derived from William Jones. If such sources do exist, then we may owe the water-leaper an apology, but at present the likelihood of a toad with wings even in the rarefied realms of Welsh mythology seems on par with that of flying pigs!

A WINGED TOAD FROM ENGLAND?

In the November 2012 issue of the periodical *Flying Snake*, British cryptozoologist Richard Muirhead reproduced the following letter, written by Thomas Flatman of Mendham, Suffolk, to his brother on 25 September 1662. Richard had unearthed it while perusing the Early Modern Letters Online database at Oxford University's Bodleian Library:

> I have iust [just] leysure enough to answere that part of yours wch [which] concernes the newes of the Serpent- amongst us, I have not seene it myselfe but can name you 20 yt [yet] have all agreeing punctually in the relacon [relation] & descripcon of ye same; tis above a yard and an halfe long an head like a toade but very large a yellowish ring about ye neck 2 wings as broad as a mans hand like a Batts 4 yellowe short leggs like a ducke as bigg as a lusty mans Thigh the Belly yellowe speckled with blacke spotts, head and back all covered with thick scales wch shine in the sunne reflect all manner of coullers hee was seen eating a water henn is most often seene before sunn rise in the morning and about noone when the Sunne shines bright and hott. Heere is one affirmes that hee surprised the Serpent one morning and being in a place where hee could not retreate hee ris: & sprung att ye man but mis't him...

Despite the fact that this creature was referred to by Flatman as a serpent, and that its yellow torque is reminiscent of that of the common grass snake *Natrix natrix*, its wings and especially its four short legs evidently rule out any ophidian identity for this mystery beast. Yet who has ever heard of a toad with wings, or even one with scales for that matter?

Bizarrely, the known animals that Flatman's bat-winged enigma most nearly recalls are the famous *Draco* gliding lizards or so-called flying dragons, which possess a pair of extendable wing-like gliding membranes, but these are endemic to Asia. Could a living specimen have been brought back to England at some time by a traveller or travelling sideshow and had later escaped? All extremely speculative, but unless Flatman's letter was a hoax, this identity does at least offer a vaguely plausible solution to what is otherwise a seemingly irresolvable riddle.

A COLONY OF WINGED TOADS IN FRANCE?

I am greatly indebted to cryptozoological correspondent Raphaël Marlière for bringing the following case to my attention. In an article from 1990 in the periodical *Communications*, French cryptozoologist Jean-Jacques Barloy presented an equally curious report of supposed winged toads. It consisted of a letter written to him on 26 August 1985 by correspondent Marcel Buisson of Alençon.

In this letter, Buisson claimed that his father had told him that during the years 1916-1921, the park of a castle near Fresnay-sur-Sarthe in northwestern France had harboured about 30 flying toads. These creatures were allegedly identical in general appearance to common toads except for one remarkable extra feature – a pair of small membranous wings similar to those of bats. They would fly away if anyone approached to within 3 ft of them, but as their flight was heavy, it never exceeded 30 ft before they came back down to the ground. They tended to be seen under poplar trees, along the main driveway of the park. One specimen was supposedly killed by Buisson's father, but was not preserved because he had no interest in it.

Not surprisingly, Barloy was thoroughly perplexed by this report. Indeed, apart from speculating as to whether they represented either some freak, teratological toad variety, or, even more exotically, a released colony of Asian *Rhacophorus* gliding frogs, he was unable to offer any reasonable explanation. However, the gliding frogs achieve their airborne state not via wing-like flank membranes like *Draco* gliding lizards, but instead by extensile membranes between their toes. As for any freak toads that had somehow developed wings: these would surely be much too precious to be allowed simply to live unfettered and unprotected in the park, and thereby take their chances against the ever-present threat of predation from foxes, cats, and other carnivorous animals. What a tragedy that the killed specimen was not preserved – and what happened to these astonishing creatures after 1921? Once again, we have a reputed record of winged toads that defies any satisfactory resolution.

WINGED TOADS – OR FLYING FROGS - IN INDONESIA?

Indonesia has a long folkloric tradition with regard to winged anurans. According to legend, a flying frog is a very special frog that after going through the normal metamorphosis from a water-dwelling tadpole to an adult frog that can live either in the water or on land, then progresses through an additional metamorphic stage, in which it grows wings and thus becomes a frog that can live in the water, on land, or in the air. Consequently, this extraordinary creature symbolises change and transformation on a spiritual level, and is also a bringer of good fortune, because it can mediate between the elements of water, earth, and air.

The Indonesian island of Bali in particular is famous for its beautiful, multi-coloured wooden

carvings of flying frogs (although some more closely resemble toads than frogs), which are frequently suspended over the cribs of babies to protect them. They are also hung as mobiles over the doorways of the homes of menopausal women, who, by progressing from their years as care-givers to the more liberating years as wise women, are thereby transforming from the physical to the spiritual.

Engraving from 1904 of a Bornean gliding frog *Rhacophorus pardalis*

But from where has the notion of flying, winged frogs in Indonesia originated? Again, the most obvious sources of inspiration, and much more likely too in this instance than in the case of the French flying toads, are the *Rhacophorus* gliding frogs, first brought to scientific attention by evolutionist Alfred Russel Wallace during the mid-1800s (see this chapter's opening quote). Sometimes referred to as flying frogs themselves, in reality they are only capable of passive gliding rather than actively-powered flight, and as already noted they are not equipped with wing-like gliding membranes. Instead, they glide by spreading out their legs and extending their greatly-enlarged interdigital membranes when leaping from trees in their native rainforest domain.

Nevertheless, at least a dozen species of gliding frog occur in Indonesia, including Boulenger's *R. modestus*, the Sumatran *R. poecilonotus*, Javan *R. margaritifer*, Sulawesi *R. edentulus*, and Wallace's *R. nigropalmatus*. Consequently, the sight of such remarkable amphibians gliding through the air may well have been sufficient to inspire fables and superstitions featuring greatly-elaborated, exaggerated versions sporting bona fide wings.

A much less likely but not impossible alternative possibility is that there is – or once was – an undiscovered creature in Indonesia that at least superficially resembled a frog (or toad) with wings or wing-like gliding membranes.

FLYING TOADS – AT A MARKET NEAR YOU!
Finally: Although regrettably not of the animate variety, flying toads and frogs with wings are

Balinese wooden mobile of a flying toad (Dr Karl Shuker)

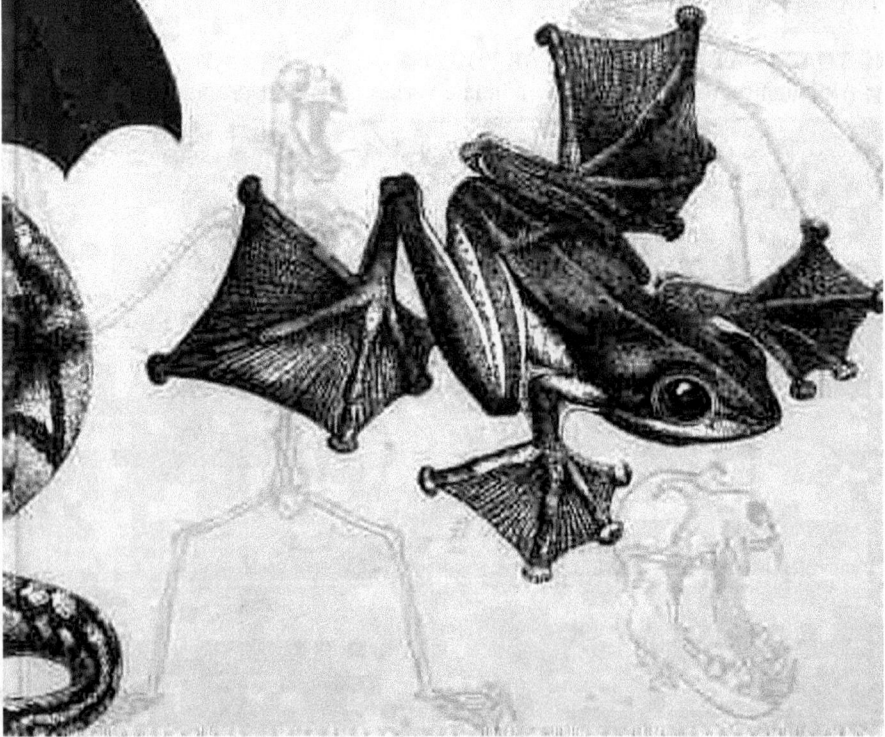

The Czech edition of my book *From Flying Toads To Snakes With Wings*, featuring on its front cover an engraving from 1896 of a Javan gliding frog *Rhacophorus margaritifer* (Dr Karl Shuker)

**My flying toad plush toy, rescued and repaired
(Dr Karl Shuker)**

by no means as difficult to encounter as you may suppose.

As someone who has authored a book entitled *From Flying Toads To Snakes With Wings*, I was never going to abandon a damaged flying toad when I encountered it several years ago at a bric-a-brac market. As seen here, it is an enchanting plush toy with flexible wings, but, sadly, there was a small hole in its back through which some plush had begun to protrude. Thanks to the deft application of some fabric glue, however, the hole is

**One of the three very large Balinese winged frog carvings spotted by me at a
car boot sale (Dr Karl Shuker)**

Balinese wooden mobile of a flying frog (Dr Karl Shuker)

no more, and said flying toad now resides in fully-restored comfort in my study.

So too does a second flying toad, this time a Balinese specimen carved out of wood, brightly painted, and with removable wings. One Sunday a couple of summers ago, I had spent a pleasant morning wandering around a very large outdoor car boot sale not too far from my home, but by the time that I'd walked down all of the aisles twice it was drawing to a close, with most of the sellers packing their unsold items away, ready to depart. Nevertheless, I decided to take a final look at the few stalls that were still selling, but when I came to one of them, I realised that it included a side-stand I hadn't noticed during either of my previous visits to it that morning. And there, hanging forlornly from that hitherto-overlooked side-stand, was the charming carved mobile pictured on p.137, which I swiftly purchased for under £5.

Some weeks earlier, I had seen a trio of extremely large Balinese winged frog carvings – I hesitate to call them flying frogs due to their substantial size (the biggest stood about 3 ft tall) – propped up against a car wheel and lying on the ground at a different car boot sale. Yet despite their poor state of preservation (their paint was peeling badly, and chunks of their wood had begun to split), their seller was asking £30 for each one. So, regrettably, these carvings did not find themselves added to my collection. Instead, I contented myself with snapping a few photographs of them, including those on pp. 139 and 140:

As for the living flying toads documented in this chapter: until – if ever – a specimen of any of them is obtained for scientific examination, they are destined to remain unsolved anomalies in the vast chronicles of cryptozoology.

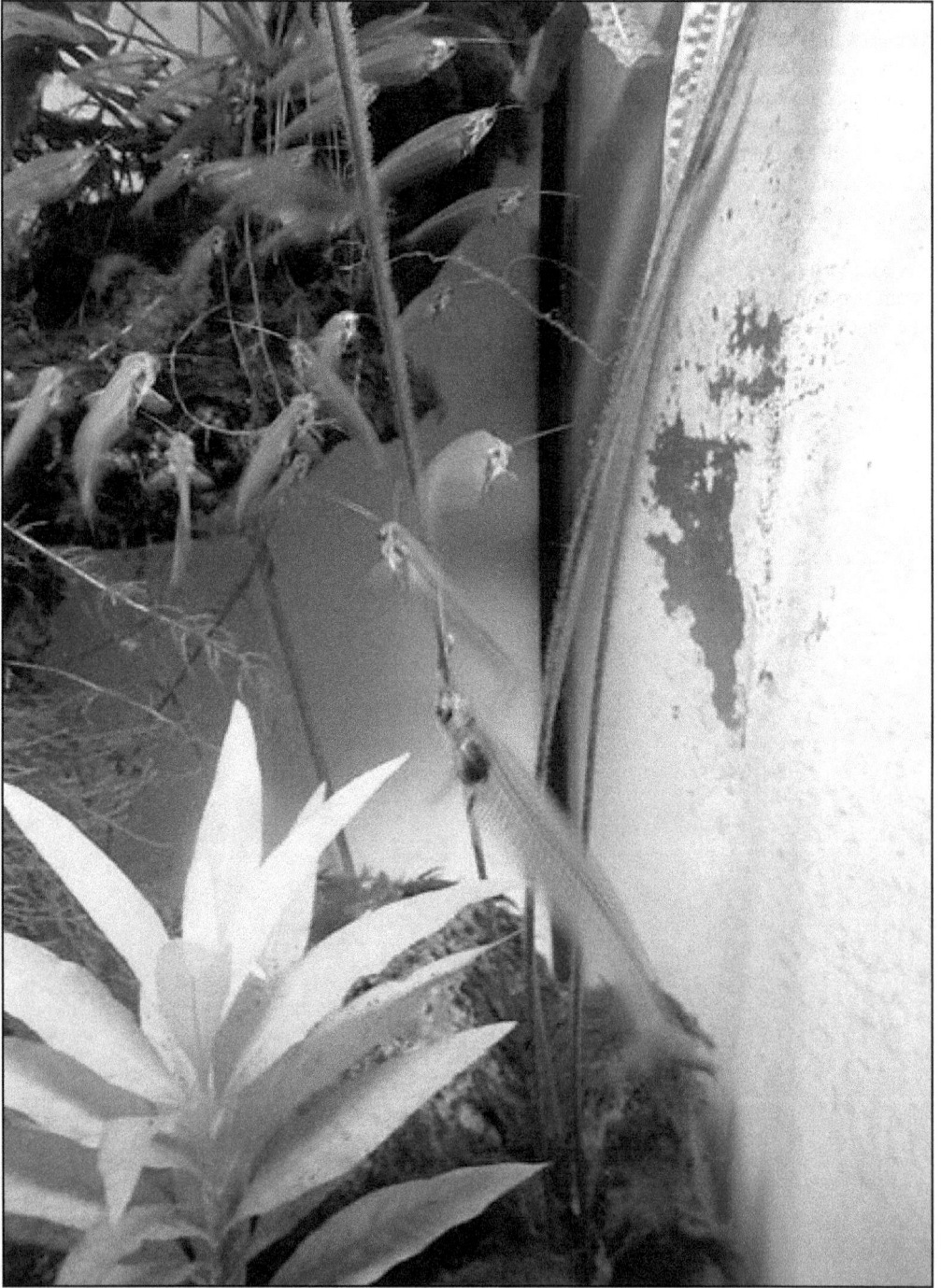

Glass catfishes *Kryptopterus bicirrhis* (Wikimedia Commons)

Chapter 14:
LOOK OUT FOR THE INVISIBLE CATFISH!

The other day in its gloomy lair
I saw a fish that wasn't there.
It wasn't there again today,
Oh, how I wish he'd come and stay.

Ivan T. Sanderson – *Caribbean Treasure*

There are several species of fish familiar to the tropical freshwater aquarist that are virtually transparent. These include the x-ray fish *Pristella maxillaris* (a species of tetra), the glass catfish *Kryptopterus bicirrhis*, and the ghost catfish *K. minor*. But what about an entirely transparent, invisible fish?

By definition, no-one has seen such a creature, because if they have done so, it couldn't have been invisible – or could it? During some cryptozoological researches, I uncovered an intriguing account of an allegedly invisible species of catfish, encountered and reported first-hand by a famous zoologist. As will be revealed here, however, upon further investigation it turned out to be something far removed indeed from its original description.

A SEE-THROUGH SECRET FROM THE SEYCHELLES

As also noted elsewhere in this book, serendipity plays a not-inconsiderable part in cryptozoology, at least in my experience, because as has happened on a number of other occasions too, I came upon this particular case while investigating a totally separate one. The latter, unrelated case had been brought to my attention by Gerald L. Wood, the author of all three editions of the exhaustively-researched, still-definitive book on zoological superlatives, *The Guinness Book of Animal Facts and Feats*, and who was also a longstanding friend of mine. In a letter to me of 1 July 1990 that referred to a number of different mystery animals, Gerald included the following brief but tantalising enquiry:

> Do you know anything about a new species of fish that can make itself
> invisible? Discovered near coral reefs off the Seychelles in the Indian

Ocean this mysterious creature turns from black to grey before 'vanishing'! Apparently a pair sell for £15,000.

I had certainly never heard of it before, but knowing Gerald well, I had no doubt that this was a serious request on his part, not a joke; if he was asking me for information concerning such a fish, then he definitely believed that it existed. So I promised him that I'd look into it, and get back to him with any news that I may find. Tragically, however, this was not to be, because only a short time later Gerald died suddenly. And despite my efforts, I never did succeed in adding any details to those scant ones supplied by him.

This episode took place several years before the internet became an unrivalled source of instantly-accessible data. More recently, therefore, after recalling Gerald's invisible mystery fish and re-reading his letter, I pursued it again, but this time online, to see if anyone else had ever reported such a remarkable creature. Sadly, I still failed to elicit any information concerning it, but I did learn about what sounded like a bona fide invisible catfish, indigenous to a specific freshwater cave pool on the West Indian island of Trinidad.

IVAN SANDERSON'S NOT-SO-CRYPTIC CARIBBEAN CATFISH
My information source was a passage of text from a book entitled *Caribbean Treasure*, first published in 1939 (thanks to Cameron A. McCormick on Facebook for providing me with a copy of the passage). It was written by Ivan T. Sanderson (1911-1973) - a Scottish-born American zoologist who was also an animal collector, zoo founder, prolific nature-travel writer, and notable television personality in the States (in many ways, therefore, a direct counterpart to Britain's own Gerald Durrell).

Sanderson noted in his book that he had been conducting a field trip to Trinidad's Northern

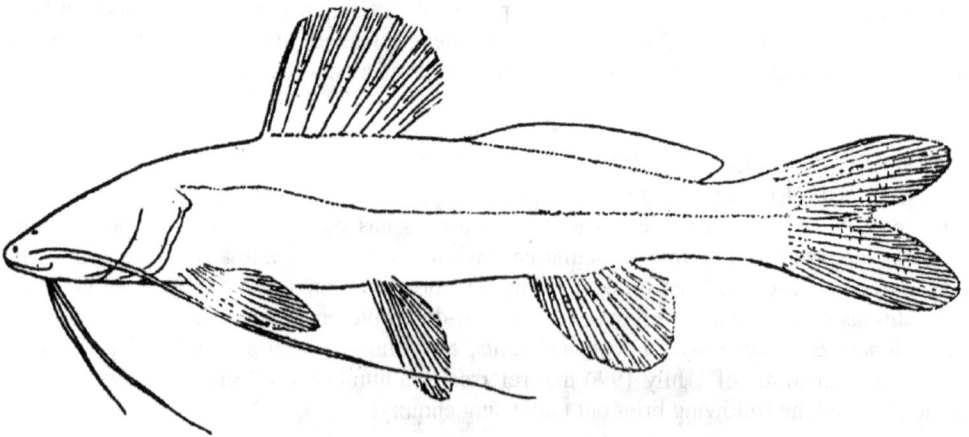

Line drawing of *Caecorhamdia urichi*

Range when he was informed by his local guides that a certain pool at the foot of the first vertical drop of Oropuche (aka Cumaca) Cave was the only known habitat of a rare, unique species of catfish that was so colourless and transparent that it could only be detected by observing its shadow passing across the bottom of the pool. Sanderson identified this elusive species as *Caecorhamdia urichi*, and stated that it was totally blind. Due to its invisible nature, no specimen was captured by Sanderson or his helpers, even when using a torch beam - hoping to illuminate it somehow.

That, at least, was Sanderson's claim concerning this species. The reality, however, as I discovered when seeking out more information regarding it, is very different indeed. It was first brought to scientific attention in July 1924, when Trinidad-born naturalist Friederick W. Urich sent a specimen to London's Natural History Museum. After studying it, in October 1926 museum ichthyologist John R. Norman formally described and named its species *Caecorhamdia urichi*, in honour of Urich.

During the mid-1950s, six more specimens were collected in its cave pool by Prof. Julian S. Kenny, the foremost expert on Trinidadian freshwater fishes at that time. After studying them in aquaria maintained at his home, Prof. Kenny concluded that they were not a valid species in their own right but merely a troglobite (cave-dwelling) variety of *Rhamdia quelen* – a species of three-barbelled catfish common in rivers throughout Trinidad.

Moreover, these six specimens varied greatly in colour, from dark grey-charcoal to pale pinkish-white. Yet all were readily visible, being quite thick in shape (as opposed to the extremely thin, flattened shape that one would expect for a reputedly transparent fish), and had therefore been easily captured. And whereas the pale specimens were indeed eyeless, the darker ones possessed small but well-formed eyes. Clearly, therefore, Sanderson's description of this catfish form was incorrect on a number of crucial counts. In addition, I remain baffled at how anything supposedly invisible by being totally transparent is able to cast a shadow anyway.

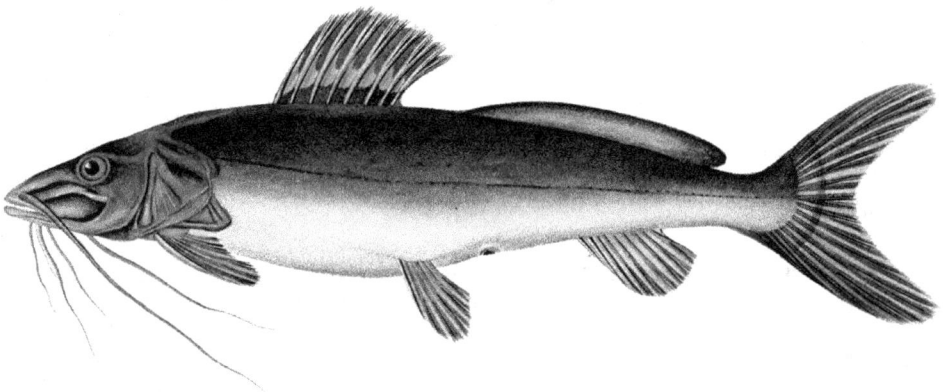

Rhamdia quelen

In April 1966, the plot thickened even further, when Dr G.F. Mees, a catfish expert from the Netherlands, tried to catch some specimens in their cave pool. In contrast to Kenny's experience, they proved very difficult to capture, and when he finally did procure three specimens, Dr Mees was surprised to discover that two of these were normal-coloured, eyed specimens of *R. quelen*, and the third, although eyeless, was also normal-coloured.

In October 2000, Dr Aldemaro Romero and Joel E. Creswell published a short article in *National Speleological Society News* concerning this fish and their January 2000 visit to its pool, where they observed dozens of specimens. Not one of them, however, was eyeless or of pale, depigmented colouration. On the contrary, their eyes each appeared to possess a tapetum lucidum, making them flash when lit by torchlight. Romero and Creswell concluded that although there may well have originally been pale, eyeless specimens here, they were probably rendered extinct following an influx of normal-coloured, eyed specimens from a stream that had invaded their cave.

There is a notable precedent for this hypothesis. A population of the Mexican cave tetra *Astyanax mexicanus* was documented in 1983 that had originally consisted of pale, eyeless specimens, but these had been wiped out in under 50 years following an influx of normal-coloured, eyed specimens from a river close by.

Today, *C. urichi* is treated merely as a synonym of *R. quelen*, and the allegedly invisible nature of its former representatives as claimed by Sanderson (in what was ultimately dismissed by critics as an exercise in 'creative description') has been wholly disproved. A sad but perhaps fitting conclusion to this remarkable case – an invisible catfish that was not invisible at all in real terms, but was finally rendered so via taxonomy.

DON'T OVERLOOK THE BRAZILIAN INVISIBLE FISH!
No coverage of invisible fishes could be complete without mentioning the infamous Brazilian invisible fish. Once a staple exhibit in any travelling sideshow or display of curiosities, it was normally housed in a large water-filled goldfish bowl, and the viewing public were invited to peer closely at the bowl in case they could discern this rare, elusive species. Some observers couldn't spy it, which is not really surprising, because except for the water there was nothing whatsoever in the bowl!

The Brazilian invisible fish was, of course, a hoax. It first attracted notable attention when Harry Reichenbach (1882-1931), an American publicist, used this scam in order to attract potential customers to a poor woman's restaurant, by placing the bowl and a big sign advertising it in the store's window.

Amazingly, however, there would always be those who were adamant that they had definitely seen something move inside the bowl - and sometimes that was actually true. This was because Reichenbach would strategically place a small electric fan out of sight but near enough to the bowl to create a faintly visible ripple passing through the water. All of which goes to prove that just as there are none so blind as those who do *not* want to see, equally there are none so perceptive as those who *do* want to see. A noteworthy cryptozoological caveat?

Chapter 15:
TWO SEA MONSTERS FOR THE PRICE OF ONE!

> Men really do need sea-monsters in their personal oceans...An ocean without its unnamed monsters would be like a completely dreamless sleep.

John Steinbeck - *The Log From the Sea of Cortez*

Sea monsters can be very deceiving, even when dead. For example, it is well known (especially in cryptozoological circles) that the decomposing carcase of a beached basking shark *Cetorhinus maximus* often transforms very dramatically, and deceptively, to yield what on first sight looks remarkably like a long-necked, four-flippered, slender-tailed, hairy plesiosaur-like creature. This is the so-called pseudo-plesiosaur effect - in which the jaw and sizeable gill apparatus fall away, revealing a lengthy portion of vertebral column that superficially resembles an elongate neck; coupled with the shark's dried-out pectoral and pelvic fins looking like flippers; its lower tail fin dropping off to yield what ostensibly seems to be a long slender tail; and its skin's exposed collagenous connective fibres gaining the appearance of thick fur.

Similarly, when a sperm whale *Physeter macrocephalus* dies at sea and its carcase gradually rots, its heavy skull and skeleton eventually sink down to the ocean floor, but sometimes a very sizeable skin-sac of rotting blubber, surfaced externally with exposed connective tissue fibres, will remain afloat - encasing a thick matrix of collagen and often not only the substantial spermaceti organ too but also a few isolated ribs with fibrous flesh still attached. If subsequently washed ashore, becoming what is popularly dubbed a globster, this hairy, bulky, gelatinous mass, with the ribs protruding like tentacles, is sometimes mistaken for the mortal remains of a gargantuan octopus – an extraordinary metamorphosis just as radical as the pseudo-plesiosaur effect, and one that a few years ago I dubbed the quasi-octopus effect (see my book *Extraordinary Animals Revisited*, 2007).

Obviously, however, as a pseudo-plesiosaur only arises with decomposing sharks whereas a quasi-octopus only occurs with decomposing whales, there is no mechanism by which both of these cryptozoological artefacts - these charlatan sea monsters - could result from the same carcase. Or at least that is what I had always assumed – until the following case was brought to my attention, and which was hitherto undocumented until I wrote it up in a *Fortean Times* article (March 2012).

It was on 6 February 1996 that Roger C. Reeves from Brisbane in Queensland, Australia, wrote a short letter to me informing me of what he referred to as a mysterious rotting sea creature that had been seen – and photographed - lying on a beach in Kent, England, in 1976 by his secretary, Juliet Lilienthal, who lived there at that time. Roger kindly provided me with his secretary's current address, and after writing to her I received four excellent close-up photos of the carcase, three of which are reproduced here (the fourth was merely a paler version of Photo #1).

With the photos (which she kindly permitted me to retain and use for my researches if I so chose), Juliet enclosed the following short letter:

> The enclosed photographs were taken in 1976 and I was wondering if the creature was a form of shark. It was washed up on the beach, it had a form of scales on the body and seemed like feathers on the neck – the head was only bone and gristle, the tail was long, similar to a crocodile with (which seemed) elephant hair at the base of the tail. It must have been damaged by a boat because the lungs were spread out on the beach (as seen in photo). It had flippers and feet, and was pregnant.

How a basking shark carcase decomposes into a pseudo-plesiosaur (Markus Bühler/*Journal of Cryptozoology*)

Combined with the above description, the photographs depict a classic pseudo-plesiosaur, from the cartilaginous (gristle) skull or chondrocranium possessed by sharks, and the long neck created by the jaw and gill apparatus falling away, to the flippers, and the long crocodile-like tail resulting from the breaking off during decomposition of the lower fin, leaving behind only the upper fin (into which a shark's backbone runs).

The scales were the rough, tooth-like dermal denticles borne in the skin of sharks, and the neck 'feathers' were strands of exposed connective tissue, as were the 'elephant hair' at the tail base. So far, so good.

As can be seen in Photos #1 and #2, however, also present was what looked for all the world like a mini-globster, lying on the beach a little way apart from the rest of the carcase (i.e. the

Photograph #1 - Kent sea monster carcase 1976, with pseudo-plesiosaur portion above mini-globster portion (Juliet Lilienthal).

Photograph #2 - Kent sea monster carcase 1976, with pseudo-plesiosaur portion on the left (head lowermost) and mini-globster on the right (Juliet Lilienthal).

Photograph #3 - Kent sea monster carcase 1976, showing close-up of pseudo-plesiosaur portion's body and the tissue section linking it to the mini-globster and containing the apparent gill arches (Juliet Lilienthal).

Photograph #3a - Photo #3 of Kent sea monster carcase 1976, with the apparent gill arches outlined by Markus Hemmler (Juliet Lilienthal/Markus Hemmler).

A living, pre-pseudo-plesiosaur basking shark!

pseudo-plesiosaur portion), and possibly placed there specifically by one or more of the interested onlookers (of which there were many, judging from the photos), but still physically linked to it by tissue. In her note, Juliet presumed that this peculiar object was the creature's lungs, but as the creature was a shark it obviously did not possess lungs, respiring via gills instead. So just what was the mini-globster?

As it was part of the shark carcase, it was evidently not a true globster, i.e. a quasi-octopus, composed of whale blubber and collagen. Instead, it was undoubtedly an organ of some kind, but in view of its very large size in relation to the main, pseudo-plesiosaur portion of the carcase it was no ordinary one.

In Photos 2# and 3#, the mini-globster appeared to be connected to the pseudo-plesiosaur portion via a series of white, bony-looking arches, which I assumed were the gill arches component of the gill apparatus. As for the mini-globster itself, the only organ in sharks attaining such a size is the liver. In some species, this massive, bilobate, oil-storing mass can account for as much as 25 per cent of the shark's total body weight, and can occupy up to 90 per cent of the total space present within its body cavity! Moreover, with regard to the basking shark, the liver is so substantial that in a 2072-lb individual, it can yield as much as 549 gallons of shark liver oil!

Keen to receive some independent opinions, however, I showed the photos to two colleagues who share my interest in 'sea monster' carcases. One was Markus Hemmler, a German cryptozoologist with whom in September 2010 I had already successfully uncovered the identity of a uniquely perplexing sea monster – the enigmatic Trunko (see my book *Mirabilis*, 2013, for full details). The other was British palaeontologist Dr Darren Naish, who has surveyed some classic sea monster corpses in his Tetrapod Zoology blog (at http://blogs.scientificamerican.com/tetrapod-zoology); as indeed has Markus in his own Kryptozoologie-Online blog (at http://www.kryptozoologie-online.de). What were their views?

Markus agreed that the white arches were probably the gill arches, and he prepared a version of Photo #3 in which he outlined these (Photo #3a here). As for the mini-globster itself, he wondered whether this may constitute some remains from the shark's pectoral girdle.

Conversely, noting that the mini-globster seemed to lack any vertical bars, which should still be present if it was part of the gill apparatus, and also that it looked much too solid in form to be the latter, Darren leaned towards it being the liver, confirming that the liver of sharks is indeed huge, extending for much of the body's length. Moreover, the texture of the mini-globster is somewhat liverish in appearance.

Although I wrote back to Juliet requesting any further information that she could offer me, particularly in relation to the precise location in Kent where the carcase had turned up, I didn't receive any further response from her, and I have no knowledge of whether any samples were taken from it for scientific analysis.

Even so, the photos (which are among the best that I have seen of a supposed sea monster carcase) and her written description are sufficiently informative for me to state with an unexpected degree of confidence when dealing with such notoriously ambiguous specimens as sea monster carcases that the Kent 'two for the price of one' example from 1976 was assuredly a highly decomposed shark. In view of its large size (using the onlookers surrounding it as a scale), it was probably a basking shark, with the mini-globster most likely a portion of its gill apparatus or (the identity I personally favour) the shark's liver.

If any of this present book's readers happen to have been among those onlookers who in 1976 viewed the Kent sea monster carcase documented here, and could provide additional information concerning it, I would love to receive your comments and information.

Chapter 16:
WHEN THE SHROPSHIRE MAMMOTHS CAME TO TOWN

The bones were as fresh as if they had just come out of a butcher's shop.

Dr John Baker - Liverpool University Veterinary Pathology Department (to whom the bones of the Shropshire mammoths were sent for x-raying and state of health assessment)

The important point to note is that ordinary people with a sense of curiosity and concern can still make major contributions to our knowledge.

Geoff McCabe, County Museums and Arts Officer, Shropshire County Council – speaking about the discovery of the Shropshire mammoths

[I wrote the original version of this chapter in March 1987, and it appeared later that year as an article - one of my earliest publications - in a now long-defunct British monthly magazine called The *Unknown*. In order to maintain its then-current, now-historical flavour, I am republishing it here in largely unchanged form (except where newer information and discoveries have required some minor updating of material).]

I n February 2014, I visited the Shropshire Hills Discovery Centre at Craven Arms, near Ludlow, Shropshire, ensconced in some of England's most beautiful countryside, in order to see the life-sized skeleton replica of a certain, very special Ice Age mammal – and, in so doing, revisiting an extraordinary discovery that I first documented way back in 1987. Allow me to explain.

One of the most remarkable yet unexpected palaeontological finds of modern times in England took place in the county of Shropshire, and involved a discovery of truly mammoth proportions.

The saga began inauspiciously at the end of September 1986 during a session of excavations by contractors working in an ARC Western-owned sand and gravel quarry at Condover, a small village just north of Shrewsbury in Shropshire. Quarryman Maurice Baddeley was using a dragline to scoop up clay and peat sediment from the quarry's upper surface in order to reach the gravel underneath, piling the removed sediment into a towering pinnacle for subsequent levelling. During this activity, his dragline's bucket drew up from a muddy pond a long, stiff object that Mr Baddeley initially dismissed as a metal or wooden post, probably a telegraph pole, and tipped onto the sediment pile. Upon later, closer observation, however, just prior to the pile being demolished, he realised that this 'pole' was actually a gigantic bone – measuring 4-5 ft long!

At this same time, Eve and Glyn Roberts of nearby Bayston Hill were walking their dogs here and saw the bone. Realising that it might be something important, Eve rapidly telephoned the Shropshire Museums Service, and relayed what they had seen to the County Museums Officer, Geoff McCabe, who promptly sent out a team to investigate. To their great surprise and delight, the team discovered that the intriguing object was nothing less than a limb bone from a woolly mammoth *Mammuthus primigenius* – that hairy elephantine epitome of the Ice Ages.

Naturally the scientists immediately combined forces with the contractors to monitor future digging in the hope of disinterring further remains - with deserved success. For during the next week, 18 more specimens were obtained, including various vertebrae and a jawbone bearing two enormous teeth.

Fossil remains, even when as massive as those of mammoths, are unexpectedly fragile when unearthed. Hence to ensure their continued survival, the precious Shropshire specimens were swiftly transferred to nearby Ludlow Museum, where they could not only be more precisely identified and age-determined but also be carefully cleaned of debris, shielded from harmful sunlight, and allowed to dry very slowly to prevent distortion.

Meanwhile, the regular media reports concerning the mammoth's discovery, as featured in the *Shropshire Star* newspaper in particular, had incited very considerable public interest - resulting in the brief unveiling of these remains for a press conference and photo-session held at the local Acton Scott Farm Museum on 7 October.

FORMAL SCIENTIFIC EXCAVATION
Among the scientific representatives present at the conference was Dr Russell Coope - Reader in Palaeontological Sciences at the University of Birmingham. On the morning of 9 October (and subsequently working in conjunction with mammoth expert Dr Adrian Lister of the University of Cambridge), Dr Coope led the first formal scientific excavation at the quarry seeking more mammoth remains. Moreover, news of this most significant search had already travelled beyond Shropshire, because the BBC's long-running children's television show *Blue Peter* was represented on site by presenter Mark Curry and an attendant film crew, recording the excavation for inclusion within a future episode (which was screened on 30 October).

The four-day dig (financed by ARC and the Shropshire County Council) brought together a team of scientists from the University of Birmingham and the Shropshire Museums Service plus numerous enthusiastic local volunteers. Their principal focus of attention was the 20-ft-high sediment pile already hewn out of the quarry by the draglines, because it was this sediment that had originally contained the mammoth's skeleton - and from which, therefore, the team hoped to disinter and disentangle it, piece by piece.

AN OVERWHELMING SUCCESS

By the close of Day 1, even the most optimistic expectations had been exceeded, because a tally of over 50 specimens - ranging from tiny wafers of tusk fragments to entire limb bones - had been unearthed! These were lightened by removing loose debris, and each specimen was then delicately packed separately within an opaque, fully-labelled bag for direct transportation to Ludlow Museum for identification and preservation. Deer and insect remains, plus pollen samples, were also collected.

The search ended on 13 October, and proved to have been an overwhelming success, because with more than 200 separate bags of fossilised remains, it seemed certain that almost the entire mammoth skeleton had been obtained. Pride of place within the collection, however, was surely the pelvic girdle, because part of it was obtained intact as a single, massive, and substantially heavy portion bearing one complete acetabulum (the socket for femur articulation) and obturator foramen (a large gap between the pubis and ischium bones on each side of the pelvic girdle in mammals). Nevertheless, there was even more exciting news to be disclosed.

NOT ONE, BUT FIVE!

Put quite simply, scientific examination of the collection obtained at that point (as well as during three subsequent excavations, the last one spanning 15 June to 3 July 1987) ultimately revealed the presence of not one but *five* mammoths! One was an adult (originally thought to be a female, but confirmed by Russian scientist Vac Garutt at the Leningrad Museum of Science in 1988 to be a male), believed to have been 30-32 years old when it died. The other four were juveniles. Three were each represented by a largely complete lower jaw and various other remains. Two of these latter three juveniles were 3-4 years old, and appeared to be of opposite sexes. The third was larger, and was aged 5-6 years old. One of the 3-4-year-olds was found during the first dig, as was the 5-6-year-old, whereas the other 3-4-year-old came to light during the summer 1987 dig, as did the fourth juvenile, thought to be 4-5 years old and represented by a single rib.

Prior to continued scrutiny, however, it was imperative that their fossilised remains be cleaned thoroughly to remove as much tenacious debris as possible. So to ensure effective washing, an outdoor area normally reserved for the cleaning of public transport vehicles was utilised! Not surprisingly, on 1 November this unusual event attracted a large crowd of spectators.

The scientific team estimated that approximately 80 per cent of the total skeletal content of the mammoths had been obtained during the recent excavation. Nevertheless, one major item was still missing - the adult mammoth's skull. Undaunted, the team decided to instigate a second

A selection of my woolly mammoth memorabilia (Dr Karl Shuker)

search, and once again, following a public appeal for local volunteers, a sizeable party was assembled, wielding a formidable armoury of shovels and spades. Yet sadly, despite a most valiant and determined effort sustained throughout the weekend of 15-16 November, the skull was not located, although several additional minor bones were unearthed. A third excavation took place not long afterwards, with a fourth, final dig taking place the following summer, but the skull was never found. As suggested by Geoffrey McCabe, it may have been removed soon after the adult's death by human contemporaries.

FUTURE DESTINATION
Even without the skull, however, the Shropshire specimens still constituted one of the most comprehensive collections of mammoth skeletons ever discovered. Indeed, the County Museums Service hoped to retain them to form the centre-piece of an extensive educational exhibition, depicting the appearance of Shropshire during the Ice Ages when inhabited by mammoths.

In turn, this would also greatly benefit local tourism. Conversely, in view of their national scientific significance, it was equally possible that they may be taken for permanent display in London. Thus in December 1986, a local conference was held to discuss the mammoths' future destination, attended by Shropshire Council members, ARC

Remains from the Shropshire mammoths (courtesy of the *Shropshire Star*)

representatives, and Drs Coope and Lister.

To the Shropshire community's delight, it was decided that the collection should be retained locally, for the planned Ice Age exhibition. Furthermore, ARC gave permission for future digging in 1987 in pursuit of any further remains (including the adult mammoth's elusive skull), and pledged financial participation in subsequent scientific studies upon the bones already obtained. Then in March 1987, the entire collection was transported to the University of Birmingham for research purposes.

The Shropshire mammoths' scientific debut - via a formal paper written by Coope and Lister and published in the scientific journal *Nature* on 3 December 1987 - was certainly one of the most thrilling episodes in British palaeontology for very many years, supplemented by continuing detailed studies. Even so, although certainly not the types of fossil to be found every day of the week, mammoth remains have been uncovered in the UK before - so why were the Shropshire mammoths of especial importance? This can be readily answered as follows.

PHENOMENAL COLLECTION
British remains of mammoths and other fossil elephants almost invariably consist of a few

Schematic representation of a woolly mammoth

bones, teeth, or fragments. Furthermore, a large proportion of these originate from the London region, although a notable find took place in Nottinghamshire during summer 1986, when two huge proboscidean (probably mammoth) limb bones were hauled up by an excavator during the construction of a car-park at Worksop's Bassetlaw Hospital. Consequently, the discovery together of a largely complete adult mammoth skeleton and no less than four partially complete juveniles is truly phenomenal.

Indeed, possibly the only British find in any way comparable to this within modern times was the unearthing in the early 1960s during excavations at an Aveley quarry in Essex of a virtually entire mammoth skeleton. Beneath this was a similarly near-complete skeleton of a straight-tusked elephant *Palaeoloxodon antiquus*, which by sheer coincidence had died on the very same spot (but undoubtedly several millennia before the mammoth - for the two species were not contemporaries). Needless to say, this unexpected but very remarkable find quickly brought a team from London's Natural History Museum to the site to remove the collection for preservation and study.

My copy of the official information leaflet from the Shropshire Mammoth Exhibition at Cosford Aerospace Museum, 1 April to 30 October 1988 (Dr Karl Shuker)

YOUNGEST BRITISH MAMMOTHS

During the last Ice Age (Weichsel/Würm glaciation), spanning the period 80,000-10,000 years BP (Before Present Day) when *M. primigenius* still roamed Britain, Shropshire was a birch-dominated tundra interspersed with sparse vegetation and clay-walled marsh-like pools created by melting subterranean ice left stranded by retreating glaciers. Dr Coope opined that the Shropshire mammoths may have wandered into one such pool while seeking vegetation. Although all elephants can swim, they cannot climb steep inclines such as the pool's walls. Consequently, the mammoths would have perished - a tragic end for such majestic creatures.

Coope explained that their remains were discovered between an upper layer of peat (shown to have been deposited 10,000 years ago) and a lower layer of glacial gravel (deposited 18,000 years ago) - another clue to the Shropshire mammoths' especial importance. More precise analysis of the bones themselves, via carbon-dating techniques, yielded an age of approximately 12,800 years BP. Hence, as Coope announced to the media, the Shropshire mammoths were not only the most complete but also, by around 5000 years, the youngest mammoths so far discovered in Britain and had survived beyond the coldest stage of the last Ice Age. In June 2009, Lister revealed that a new, even more accurate method of radiocarbon dating applied to the remains by researchers from London's Natural History Museum had yielded a date of 14,000 BP, but this still meant that they were Britain's youngest mammoths.

In addition, it was suggested that they might even participate at some future stage in one of the most remarkable fields of mammoth-related zoological research currently in progress - the cloning of mammoth DNA. The raw materials (muscles, soft tissues) for this revolutionary work are normally obtained from ice-entombed specimens obtained in Siberia. However, Prof. Alan Wilson of the University of California suggested that the Shropshire specimens may be sufficiently well-preserved to possess samples of soft tissue capable of being used for DNA cloning purposes, and he duly made contact with the Shropshire team to discover more concerning this exciting possibility. In short, the mammoths of Shrewsbury certainly appear set to occupy a prominent position within future scientific research for some considerable time to come.

During 1988, I visited the temporary but exciting exhibition of the Shropshire mammoth remains (which also included an excellent full-sized replica of a woolly mammoth, created by Roby Braun) that was held at Cosford Aerospace Museum, just outside Wolverhampton, from 1 April to 30 October of that year. The exhibition was subsequently staged in Derbyshire, Lancashire, and Newcastle upon Tyne, before closing in August 1991. The replica mammoth skeleton now resides within the Secret Hills exhibition at the Shropshire Hills Discovery Centre in Craven Arms, whereas the Shropshire mammoth bones are ensconced in the Ludlow Museum Resource Centre, and are recognised to be the third most complete remains of woolly mammoths to have been discovered anywhere in Europe. Not a bad outcome for a 'telegraph pole' that had been dug up by chance and then tipped unceremoniously onto a pile of sediment.

Alongside the replica skeleton of the adult Shropshire woolly mammoth at the Shropshire Hills Discovery Centre (Dr Karl Shuker)

Chapter 17:
FISHING FOR THE TRUTH
- A MONSTROUS MYSTERY FROM
LAKE ILIAMNA

Sightings of a huge creature that lives in the depths of Alaska's largest lake are so
persistent that the Alaska Department of Fish and Game keeps an open file labeled
"Iliamna Monster."

Alaska Magazine, January 1988

Almost twice as large as the entire U.S. state of Connecticut, Lake Iliamna in southwestern Alaska measures a very impressive 77 miles long, is up to 22 miles wide, has a surface area of around 1000 square miles, and boasts a maximum depth of 988 ft, making it the biggest lake in Alaska, and the eighth largest in the whole of the USA. It seems only fitting, therefore, that this monster-sized expanse of deep, freezing freshwater should also contain a mystery of truly monstrous proportions.

Derived from its own name, the still-unidentified water beasts said to inhabit this vast lake (and also the somewhat smaller Lake Clark that indirectly drains into it) are popularly dubbed Illies by Westerners, and are often claimed to be as much as 30 ft long. However, they are apparently very different in form from the many-humped and long-necked mystery beasts reported from various other stretches of inland water in North America.

Instead, they are usually described as very long and quite slender, greyish or dark in colour, and with a noticeable dorsal fin marked by a white stripe.

The Illie has long been known to the area's native Inuit/Aleut people, who refer to it as the jig-ik-nak, and state that it has been known to attack their boats. This cryptid is also familiar to the native Tlingits, who call it the gonakadet, deem it to be a fish deity, and have depicted it in pictographs along the Alaskan and British Columbian coasts.

Due to Iliamna's huge size, remote location, and sparsely-inhabited shoreline, Illie sightings have generally been made from planes flying over the lake, or from boats travelling across it, rather than from its perimeter. Consequently, far fewer contemporary sightings have been documented for this cryptid than for 'monsters' reported from more accessible lakes.

What seems to be the first well-publicised modern-day Illie sighting took place in September 1942. This was when local fishing guide/bush pilot Babe Alsworth and fisherman Bill Hammersley, flying over the lake in Alsworth's Stinson ferry plane toward the village of Iliamna, observed a school of several dozen animals in shallow water near an island in the middle of the lake.

They were described by the men as being dull aluminium in colour and estimated to measure well over 10 ft long, with broad heads, elongate bodies, vertical tails (this latter feature is characteristic of fishes; whales and other cetaceans have horizontal tails), and resembling mini-submarines in general form. Viewed from a height of only 300 ft at one point, the creatures circled for several minutes, but never surfaced, then abruptly surged away, hidden by the wave disturbance that their movements generated.

In 1945, a single creature of similar shape and colour, but estimated at around 20 ft long, was spied by U.S. government survey pilot Larry Rost as he flew over the lake at low altitude.

As evidenced by the Alsworth/Hammersley sighting and various others on file, the Illie swims just beneath the water surface, sometimes in groups, but unlike a number of other lake monsters it does not come up for air, remaining submerged, and therefore seemingly able to breathe underwater, like a fish rather than a mammal or reptile. This was confirmed by the next notable modern-day sighting, which took place in 1963 and featured a biologist from the Alaska Department of Fish and Game. While flying over the lake, he watched a 25-30-ft-long creature swimming beneath the water surface for over 10 minutes, during which time it never once surfaced.

In the 1950s, Bill Hammersley sought to obtain the ultimate proof that his 1942 sighting had been genuine - by attempting, with the assistance of three other enthusiasts, to hook and land a living Illie. They used a huge chunk of moose meat as bait, securely hooked via a thick 1-ft-long iron rod to a considerable length of stainless-steel 16-in aircraft cable. And sure enough, something beneath the water surface did indeed take the bait, but was so powerful that it snapped the steel cable!

In 1959, oil millionaire and longstanding cryptozoological adventurer Tom Slick offered a $1000 reward to anyone who could catch an Illie, but it was never claimed. He also hired Hammersley's co-eyewitness Babe Alsworth to fly over the lake on several occasions, but no Illie sightings took place.

In 1977, air-taxi pilot Tim LaPorte and his two passengers saw from the air a dark 12-14-ft-long animal whose back was just breaking the water surface. When it dived downwards, it revealed a large vertical tail.

On 27 July 1987, a 10-ft-long fish-like creature, black in colour but sporting a white stripe on its dorsal fin, was spied by Verna Kolyaha, her mother, and her sister while they were fishing from a skiff. Leaping and splashing, it swam in an almost complete circle around them. Their sighting occurred about 5 miles northwest of Pedro Bay village. A similar creature, swimming near the surface of the lake, was observed by multiple eyewitnesses from the shore and on the water a year later.

In 1989, Louise Wassillie (apt name!) watched what she considered to be a 20-ft-long fish with a long snout from her fishing boat on Lake Iliamna.

In May 2010, the highly popular television show 'River Monsters', fronted by fisherman-adventurer Jeremy Wade, screened Wade's recent attempt (unsuccessful) to capture an Illie.

And in May 2013, I received the following very intriguing report of possible Illie relevance from correspondent Lee Raiter, who lives in Anchorage, Alaska. It has never previously been published in any book:

> There are two brothers I know real well from [the city of] Nondalton and this happened to them. Six Mile Lake [situated between Lakes Clark and Iliamna] is where Nondalton sits and Lake Clark flows into it through the Iggyagak river and the brothers were fishing for what they call Fall Salmon, sockeyes late in the run, getting real red...They have done this for 40 years, Karl, they used to be commercial fishermen in Bristol Bay for years, no rookies. This day they had both of their 18-ft Lund boats and were sitting just off where the river goes into Clark, was a beautiful fall day, no wind, sunny out. Since fishing was slow they had their gillnet tied off on each end to both boats, 25-fathom net and were just laying back on the seats, enjoying the day, not anchored, just floating. All of a sudden the net starts to sink, then is out of sight and now both Lunds are being pulled backwards, 18-ft boats with 55 horse-power outboards on them. They had water coming over the back of both boats and were frantic to cut the net loose or start the motor, ended up with 3-4 inches of water in each boat after the net just popped back up. They had been pulled backwards maybe 10 yards, maybe 15 or so, and when they pulled the net out it was torn up some on the bottom of it. Spooked the brothers but good too. This happened about five years ago [c.2008] I think. We have seen fish on sonar three times now in last two years, minimum 16 ft [long] and my neighbour saw four of them running together on his sonar (much better) and says one was 22 ft long, the rest were smaller but not by much.

As can be seen from the above selection of accounts, there seems little doubt that the Illies are indeed fishes, albeit exceptionally big ones. The most popular and plausible identity, voiced by ichthyologists and locals alike, is a sturgeon, in particular the white sturgeon *Acipenser transmontanus*.

This mighty carnivorous species, occurring in both freshwater and the sea, is known to attain lengths of up to 27 ft, and is the largest species of fish in the whole of North America. It can live up to 100 years, and despite its name is greyish or greyish-brown rather than white in

Artist's reconstruction of Lake Iliamna's giant mystery fishes (William Rebsamen)

colour. It is also the Illie identity favoured by Jeremy Wade.

In addition, one Illie eyewitness, Eddie Behan, described the creature that he had seen as being approximately 20 ft long, spindle-shaped, with a fish's tail, and bearing rows of lumps on its back. This is an excellent verbal portrait of a huge sturgeon, the 'lumps' corresponding well to the characteristic, highly distinctive series of scutes that run like armour plating along the dorsal surface and also the flanks of sturgeons.

Although sturgeons have never been confirmed from Lake Iliamna, they are known from other, smaller Alaskan lakes. Consequently, it would not be implausible for this immense body of water to house such fishes too, and for them to attain record sizes here, due to this lake's huge volume and the plentiful food supply that it is known to contain. And as sturgeons are primarily bottom-dwellers, this would offer another reason why sightings are not more common. Moreover, there are some notable precedents on record regarding North American water monsters that were ultimately exposed as giant sturgeon.

For many years, a voracious cryptid had been rumoured to lurk beneath the dark waters of Lake Washington in Seattle, Washington State, where it allegedly preyed upon the lake's resident duck population. However, mainstream zoologists tended to dismiss such claims – until 5 November 1987, that is. For on that momentous day, a truly monstrous creature was

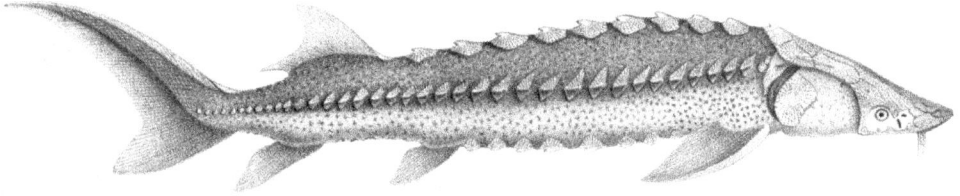

An engraving from 1868 of a white sturgeon

A 19th-Century engraving of sturgeons

indeed discovered here – a massive female sturgeon, found dead on the shore, which, when measured, proved to be an immense 11 ft long, and weighed a stupendous 900 lb. This monster in every sense of the word was estimated to be more than 80 years old. No wonder the lake's duck population had been so depleted!

More recently, during the first weekend of August 2013, history repeated itself when a family water-skiing in the northern portion of Lake Washington encountered the dead body of a very large sturgeon, floating upside-down. Brad James, a biologist at Fish and Wildlife in Vancouver, Washington State, confirmed its identity, disclosed that it measured 8 ft long, and revealed that it may have been living there for many years.

A 'monster' said to snatch away the tackle and even the fishing poles of anglers had been reported for at least 30 years at Stafford Lake, just north of San Francisco, California. In August 1984, however, this fairly small body of water needed to be drained for dam repairs – and when this task was carried out, a monster was indeed found there. It was a white sturgeon, albeit one of more modest proportions than the Lake Washington examples, but still measuring a respectable 6.5 ft long, and estimated to be 50-60 years old. It took a dozen men to capture this sturgeon alive, and it was then taken to the California Academy of Sciences' Steinhart Aquarium in San Francisco's Golden Gate Park. Sadly, however, it died just a few days later, apparently from shock.

So does a land-locked population of monster-sized white sturgeon inhabit Lakes Iliamna and Clark in Alaska? If so, they would be the most northerly representatives of their species on record, living only a few hundred miles below the Arctic Circle. With up to 20 million sockeye salmon returning to Iliamna from the sea every year, however, food would certainly not be a problem, nor would space in such an enormous volume of water (which is 15 times that of Loch Ness). Clearly it is high time that science took an interest in this too-long neglected zoological mystery, whose resolution may feature one of the greatest ichthyological discoveries of modern times.

Chapter 18:
THE GIANT RAT OF SUMATRA, AND THE MOONRAT OF MALAYA

"Matilda Briggs was not the name of a young woman, Watson," said Holmes, in a reminiscent voice. "It was a ship which is associated with the giant rat of Sumatra, a story for which the world is not yet prepared..."

Sir Arthur Conan Doyle – 'The Adventure of the Sussex Vampire',
in *The Case-Book of Sherlock Holmes* (1927)

The gymnures illustrate as well as any animals can how uneven is our knowledge. The moon rat has been known for nearly 150 years [nearly 200 years now, but the following statement is still applicable], yet our information today is little better than when Raffles was writing about it. This reflects not only the secretive habits of the animal itself but also...one other thing, that Malaya and the Malayan archipelago are rich in unusual animals so such field naturalists as have been out there have tended to give their attention to the larger and more obvious animals, such as the tapirs and rhinoceroses.

Maurice Burton and Robert Burton (eds) –
Purnell's Encyclopedia of Animal Life (6 vols)

For many people, rats of any size are the stuff of nightmares. But for me, the giant rat of Sumatra and the mighty moonrat of Malaya are totally fascinating, albeit for very different reasons.

THE GIANT RAT OF SUMATRA – ZOOLOGICAL FACT, NOT SHERLOCK HOLMESIAN FICTION

Contrary to the assumption by many aficionados of the Sherlock Holmes stories that it was wholly fictional, there really *is* a giant rat of Sumatra. Yet despite having been scientifically described as long ago as 1888 (by eminent mammalogist Oldfield Thomas of London's Natural History Museum), until as recently as the 1980s it *had* remained largely a mystery, even to zoologists.

My Sherlock Holmes toby jug confronts the giant rat of Sumatra! (Dr Karl Shuker)

In 1983, however, following an in-depth study of this noteworthy 2-ft-long rodent, Dr Guy G. Musser (Curator of Mammals at the American Museum of Natural History) and museum research student Cameron Newcomb attempted to disperse the longstanding veil of obscurity surrounding it by publishing its first full scientific description. Their paper appeared in the museum's *Bulletin*.

A very large, mountain forest-dwelling species belonging to the taxonomic family Muridae, with dense, woolly, dark-brown fur (characterised by extremely lengthy guard hairs) and powerful jaws, the Sumatran giant rat had traditionally been categorised as a typical, *Rattus* rat. After a meticulous investigation of its anatomy, however, one that surely would have met with Holmes's own approval, Musser and Newcomb recognised that its aural, nasal, and dental characteristics fully justified separation of this legendary form from the *Rattus* horde. As a result, they officially rehoused it in a new genus, *Sundamys*, along with two other Asian species.

The Sumatran giant rat is now known formally as *Sundamys infraluteus*, and is not endangered, being recorded from a sizeable area of Sumatra and also from both Malaysian and Indonesian Borneo. Its two congeners are Bartels's rat *S. maxi* and Müller's giant Sunda rat *S. muelleri*.

Formally described in 1931 as a valid species but later classified merely as a subspecies of the Sumatran giant rat until re-elevated to the rank of a species in its own right by Musser and Newcomb in their 1983 paper, Bartels's rat remains scarcely known even today. It is represented only by 21 specimens collected between 1932 and 1935 from two locations on Java by Max Bartels Jr, and is categorised as Endangered by the IUCN.

As for Müller's giant Sunda rat: originally described in 1879, this species has the widest distribution of the *Sundamys* trio, being recorded from Indonesia (including Borneo's Indonesian region and Sumatra, but not Java), Malaysia, Myanmar, the Philippines, and Thailand, and is not deemed to be endangered. It is a primarily terrestrial, lowland species.

Other extra-large species of rat belonging to the family Muridae include the two species of giant cloud rat (genus *Phloeomys*) and the four species of giant bushy-tailed cloud rat (genus

19th-Century engraving depicting three colour varieties of Schadenberg's giant bushy-tailed cloud rat *Crateromys schadenbergi*

Crateromys) endemic to the Philippines; the two giant rat species (genus *Papagomys*), one extant, one extinct, indigenous to Flores, an island in Indonesia's Lesser Sundas group; New Guinea's white-eared giant rats (genus *Hyomys*) and woolly rats (genus *Mallomys* – of its seven known species, three still await formal scientific descriptions and names); South America's woolly giant rat *Kunsia tomentosus* and fossorial giant rat *K. fronto*; and Africa's four species of giant pouched rat (genus *Cricetomys*), including most famously the Gambian giant pouched rat *C. gambianus*.

Until as recently, palaeontologically speaking, as the end of the Pleistocene epoch approximately 10,000 years ago, there was even a unique species of giant rat alive and well in the Canary Islands. Named the Tenerife giant rat *Canariomys bravoi*, it measured almost 4 ft long and weighed around 2 lb. Its fossilised remains have been found all over the island of Tenerife, but it is believed to have become extinct due to predation by cats introduced here by Tenerife's first human inhabitants.

Yet even this rangy species would have been dwarfed by the giant rats of East Timor, of which at least two distinct species – *Coryphomys buehleri* and *C. musseri* – have been described so far. Known from sub-fossils, they believed to have weighed up to 13 lb when adult, but seemingly died out 1000-2000 years ago, possibly as a result of habitat destruction caused by large-scale forest clearance for farming purposes.

As a longstanding Sherlock Holmes fan, I'm aware that although the giant rat of Sumatra's case was never documented by Dr Watson within the original, Sir Arthur Conan Doyle canon of Sherlock Holmes fiction, it has inspired several pastiches penned by other authors, yielding an extremely diverse range of identities for it.

Reconstruction of the likely appearance in life of the Tenerife giant rat, at the Museo de la Naturaleza y el Hombre, Santa Cruz, Tenerife (M0rph/Wikipedia)

A 1903 chromolithograph of the Malayan tapir - the unexpected identity of the giant rat of Sumatra in Richard L. Boyer's eponymous novel

These include such memorable candidates as a Malayan tapir *Tapirus indicus*, in Richard L. Boyer's novel *The Giant Rat of Sumatra* (1976); a monstrous mega-rat called Harat who rules a nation of sub-humans in Alan Vanneman's novel *Sherlock Holmes and the Giant Rat of Sumatra* (2003) (not to be confused with an entirely different but identically-titled novel by Paul D. Gilbert, published in 2010); and, perhaps most bizarre of all, a preternatural maritime horror, in a story penned by H.P. Lovecraft!

This mega-murid also appears in a novel featuring one of the most extraordinary literary pairings ever – Sherlock Holmes and Count Dracula! These iconic if diametrically dissimilar figures reluctantly join forces to confront a nefarious plot to destroy London using plague-bearing rats in Fred Saberhagen's *The Holmes-Dracula File* (1978), with the giant rat of Sumatra as the principal vector.

It has even left its mark in the theatre, with a number of plays having featured this rangy rodent over the years. Notable among them are the Fossick Valley Fumblers theatre group's production, 'Sherlock Holmes and the Giant Rat of Sumatra', written by Bob Bishop, and debuting at the Edinburgh Festival Fringe in August 1995; and an entirely separate but identically-titled comedy musical with music and lyrics by Jack Sharkey and book by Tim Kelly, which was first performed on 31 December 1986, by the Magnificent Moorpark Melodrama & Vaudeville Theatre, Moorpark, California, USA.

Wikipedia has a dedicated 'Giant Rat of Sumatra' page (at: http://en.wikipedia.org/wiki/Giant_Rat_of_Sumatra), containing a very lengthy listing of other literary works (plus some music and TV productions) inspired by this evocative furry entity.

All in all, a pretty impressive modern-day C.V. for a quasi-cryptid originally only mentioned very briefly in passing by a fictitious detective almost a century ago.

DEMYSTIFYING THE MOONRAT – A HAIRY HEDGEHOG THE SIZE OF A RABBIT!
As a child, animals with unusual names always held an intense fascination for me. So it was inevitable that I would want to learn more about the moonrat!

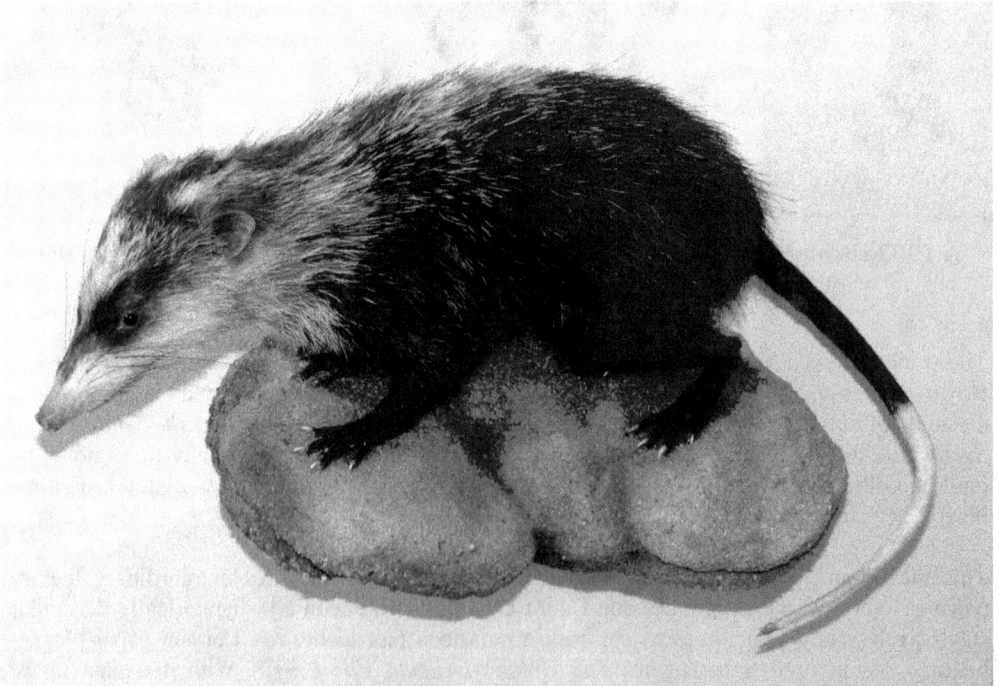

Taxiderm specimen of a moonrat (Haplochromis/Wikipedia)

174

Drawing of a white moonrat, from Robert A. Sterndale's *Natural History of the Mammalia of India and Ceylon* (1884)

Sadly, I soon discovered that in spite of its exotic appellation, this wonderful creature, known scientifically as *Echinosorex gymnura* (aka *Gymnura rafflesii* in early natural history tomes), does not actually come from the moon, but it is such an amazing-looking animal that anybody could be forgiven for wondering if it may do! In fact, the marvellous moonrat (also, though less commonly, called the bulau) is from southeastern Asia, specifically the Thai-Malay Peninsula and the large Indonesian islands of Sumatra and Borneo.

It was first brought to scientific attention in 1821, by Sir Stamford Raffles, who discovered it in Sumatra and mistakenly deemed it to be a new species of civet, duly dubbing it *Viverra gymnura*. Certain other early investigators of this mysterious mammal were equally confused by it, categorising it as a marsupial! In reality, however, the moonrat is the largest of several species of insectivore that are known as gymnures ('naked tails'). This is because their very long slender tails are almost hairless and covered in scales, rather like snakes! Equally ophidian is the loud, threatening hiss that the moonrat gives voice to if confronted by predators or by other moonrats invading its territory.

With a head and body length of 13-16 in, a tail of 8-12 in, and weighing up to 2.75 lb, the moonrat looks very like a gigantic rat - a rat wearing a black mask, and as big as a rabbit! (In Borneo, moonrats lack the mask because here they are predominantly white all over in colour, constituting a separate subspecies, *E. g. alba*, from the nominate subspecies, *E. g gymnura*, that is found everywhere else within the moonrat's distribution range.) It even occupies a similar ecological niche to true rats. Taxonomically, however, this deceptive animal is something very different, because its closest relatives are not the rodents but the hedgehogs.

For although, outwardly, it doesn't look anything like one, when its anatomy is examined the moonrat is swiftly revealed to be a kind of extra-large, long-tailed hedgehog - but one that is covered in lengthy coarse hair instead of spines. Indeed, the hairy hedgehogs is an alternative name for the gymnures.

The moonrat's external appearance has some other surprises too. Its very long, mobile snout is plentifully supplied with exceptionally lengthy, bristly whiskers - as are its eyebrows. These whiskers are extremely sensitive to touch, and probably help the moonrat to gauge in the dark whether it is thin enough to pick its way through tight crevices while active at night. Also assisting it do this is the remarkable shape of its body (which although looking very burly from the side is actually surprisingly narrow, allowing it to squeeze through gaps that seem scarcely wide enough to let it pass). So too do its very short legs.

The moonrat inhabits dense forests and swamps, usually near water. A good swimmer, it is fond of eating fishes, frogs, crabs, clams, and other aquatic creatures, as well as worms and insects. Despite its size and eyecatching appearance, however, as well as the facts that it was first documented scientifically as long ago as 1821 (by Sir Stamford Raffles no less, who named it a year later) and has a lifespan of up to five years in its native habitat, the moonrat remains one of the world's most mysterious mammals. This is because it is incredibly shy and hence is rarely seen in the wild.

One peculiar thing that we do know about its wild habits, however, is that when the female moonrat makes a nest in which she later gives birth to two babies, the fluid that she secretes from a pair of glands under her tail to mark the nest's entrance possesses a strong ammonia content and has a potent smell that closely resembles rotten onions or garlic! (The male also secretes this same fluid when marking his territory.) So even if our eyes are unsuccessful in catching sight of moonrats, our nose should have far less trouble locating their nests!

The moonrat's memorable name has assisted it in becoming an unlikely villain in a delightful children's book written by Helen Ward. Entitled *The Moonrat and the White Turtle* and originally published in 1990, it also contains Ward's beautiful full-colour illustrations. Moonrat is the greatly-feared leader of a rascally band of pirate rats, and is driven by one unquenchable ambition – to steal the moon out of the sky and add it to his vast glittering trove of ill-gotten treasure! But does he succeed with his nefarious plot, and who or what is the White Turtle? I'll leave you to discover this excellent book and find out for yourself!

There is also a rock group called The Moon Rats, but I'm unsure whether their name was gymnure-inspired, or just inspired!

Finally: all that remains to be answered is where the moonrat obtained its noteworthy name. However, this appears to be one mystery that is destined to remain unanswered, because despite

19th-Century engraving of a moonrat

considerable research, I have so far been unable to trace any confirmed origin for it.

Having said that: while I was discussing the moonrat with Australian naturalist Dr David Kirschner recently, David offered up a very thought-provoking potential explanation for its name, which he has kindly permitted me to quote:

> As for the common name, if I had to guess it would have probably originated with the Bornean subspecies, *E. gymnura alba*, as I would imagine an all white, nocturnal mammal would be quite visible on a bright moonlit night.

Indeed, taking this line of thought even further, such an animal would be so eyecatching and unearthly in appearance if viewed upon a bright moonlit night that the more fanciful of observers may even have imagined it (albeit only in jest) to have come directly from the moon itself - hence 'moonrat'!

And if anyone else also has any thoughts or knowledge regarding the derivation of 'moonrat', I'd love to hear from you!

Close-up of a gelada (Dr Karl Shuker)

Chapter 19:
NATURAL BORN KILLERS
- FLESH-EATING BABOONS
AND NEFARIOUS NANDI BEARS

One year, in a part of the Bunyoro District [of Uganda], five native children were attacked and severely bitten [by baboons] within a few days of each other; two of them died...Children, while protecting crops, are sometimes killed. The ape's [sic – baboon's] method of attack is to seize the little watchman in its arms and then virtually disembowel him with terrible, downward strokes of its muscular feet.

Colonel Charles R.S. Pitman –
A Game Warden Takes Stock

I f you think that all monkeys are cute, furry, lovable fruit-eaters - forget it, right now! Especially if you happen to be a hare, a young gazelle - or even a human baby - confronted by a troop of hungry baboons. Known to the ancients as cynocephali on account of their superficially dog-headed appearance, these large African monkeys sometimes exhibit another decidedly canine trait - a liking for meat, and are not too fussy about what they kill in order to obtain it.

Baboons typically subsist upon such mundane sustenance as grass, fruit, leaves, roots, tubers, insects, birds' eggs, and small rodents, but as a number of wildlife researchers and observers will testify, these aggressive primates are more than willing to set their sights upon larger, fleshier food sources too, should the opportunity to do so arise. American zoologists Prof. Sherwood L. Washburn and Prof. Irvin DeVore have spent many years studying baboons in several different areas of East Africa. In one reservation, baboon troops often visited the same water holes as smaller monkeys called guenons. Normally, the two species coexisted with no conflict apparent between them, the guenons moving freely among the baboons. One evening, however, the two zoologists were startled to see a baboon suddenly seize hold of one of the guenons, kill it, then promptly devour it. On another occasion, in Kenya's famous Amboseli National Park, they saw two large male baboons boldly kill and eat two newborn Thomson's gazelles, despite the efforts of their respective mothers to protect their young.

At least, however, these unfortunate victims were dead before they were consumed. In a letter to a German wildlife magazine, a South African observer, L. MacWilliam, reported how he had once seen a baboon capture a hare, then sit down and gaze for a time at its terrified victim, gripped firmly in its paws, before casually biting off one of the hare's ears and eating it, then tearing a large chunk of flesh out of its struggling victim's body and eating that, and then another chunk, and another, before finally discarding the hideously mutilated yet still-living creature and rejoining its troop nearby.

THE PUMPHOUSE GANG

The Pumphouse Gang is a troop of olive baboons *Papio anubis* (aka the green baboon) living wild on a vast cattle ranch called Kekopey in Kenya, and it became the subject of a famous, long-running study of baboon society, beginning in 1972, by primatologist Dr Shirley C. Strum - at that time a graduate student, but today a professor of anthropology at the University of California's San Diego campus and also working for part of each year in the field, continuing her baboon studies in Kenya. Prof. Strum's extensive research is documented in detail within her fascinating book, *Almost Human* (1987), which includes several observed instances of quite extraordinary predatory behaviour among the Pumphouse Gang.

19th-Century engraving of an olive baboon

It all started with a young male baboon called Rad. In the past, he had been dominated by a more tenacious hunter in the Gang, called Carl, who prevented Rad from doing much in the way of hunting himself - until Carl suffered a severe arm injury that put him out of action, thereby giving Rad free rein to his predatory passions. Soon it became a familiar sight to see Rad seek out an infant Thomson's gazelle, kill it, eat it, then return to the Gang covered in blood. At first, the other males took little notice, or, at most, were inspired to seek out similar prey for themselves, but always separately, individually. Even if several of them chased after a single gazelle, they made no attempt to combine forces and pursue it as a team - until one hunting session featuring Rad changed all that.

One day, despite chasing a baby Thomson's at full speed, Rad just couldn't get close enough to lunge at it successfully - but then the other males from the Gang appeared over the hill nearby. Seeing them, Rad deliberately chased the gazelle toward them before wearily giving up the pursuit himself to rest awhile. As he did so, one of the other baboons took over, and chased the gazelle towards another male member of the Gang, who then took up the chase in turn, in what soon became a virtual relay race, with a fresh male taking over the pursuit as soon as the previous one became tired, until finally one of the baboons chased the now-exhausted gazelle right into the arms of another one.

Clearly the success of this joint effort made a strong impression upon the Pumphouse Gang, because for a number of years afterwards Prof. Strum frequently witnessed the 'relay team' method of hunting employed by the Gang's males. Moreover, even the females in the Gang began hunting, and their youngsters learnt to eat meat by watching and imitating their mothers. A few bolder juveniles even attempted to seize baby gazelles themselves - despite the fact that they were often as big as the young baboons! Over the coming months, the Gang's rate of predation rose significantly, and instead of treating the capture of young gazelles as opportunistic, the males in the Gang now very purposefully and meticulously surveyed herds of gazelles in search of likely fawns to capture, with a hunting party travelling up to two miles and spending up to two hours before finally making a kill.

In contrast, the females never ventured from the troop in order to kill, but would very readily take meat from carcases brought back by the males. This was itself a major change of behaviour, because baboons normally never share food with one another. What was also interesting - and amusing - was the difference in priority between the females and the males in relation to meat-eating and sex. Strum would often see a female baboon staring fixedly at a tasty-looking carcase nearby while patiently tolerating a male copulating with her - clearly the baboon equivalent of lying back and thinking of England! - before determinedly heading forth for the carcase afterwards. Conversely, when offered a choice of some meat and a mate, a male baboon would briefly glance back and forth at the two options before inevitably seizing the female.

Occasionally, baboons have even been daring enough to steal human babies. One such incident took place in 1965, at an immigration camp established in Brakpan, South Africa, when a particularly bold baboon seized a baby out of its carriage in the presence of its mother, then promptly killed it by biting the poor child's head repeatedly. Uganda Protectorate game warden Colonel Charles R.S. Pitman reported that five native children were attacked in separate incidents within a single year in Uganda's Bunyoro District, and that children guarding crops were sometimes seized by a baboon and disembowelled via powerful downward kicks of its muscular clawed feet.

On the whole, however, such cases are certainly the exception rather than the rule, but nonetheless they serve as chilling reminders that the old saying about Nature being red in tooth and claw is one that we ignore or forget at our peril.

GIANT BABOONS AND THE NANDI BEAR CONNECTION

The olive baboon is a very sizeable monkey, boasting an average total length of around 48 in. This is only a little smaller than that of the largest species of baboon alive today - the gelada *Theropithecus gelada*. Native to the highlands of Ethiopia, its head-and-body length is 20-30 in, with a tail length of 12-20 in, and males average 41 lb in weight (females average 24 lb).

Unlike *Papio* baboons, however, the gelada is a vegetarian, existing as a graminivorous grazer. It is the only modern-day species in the genus *Theropithecus*, but in prehistoric times

Artistic representation of *Dinopithecus ingens* (Hodari Nundu aka Justin Case/ Deviantart)

there were other, even larger members. Of these, the most spectacular species must surely have been *T. oswaldi*, dating from the early to mid-Pleistocene of South Africa, Kenya, Tanzania, Ethiopia, Morocco, Algeria, and Spain, because this enormous baboon was as big as a gorilla. Mercifully, however, just like the gelada, it was strictly herbivorous.

But not all giant prehistoric baboons were plant-eaters. Bearing in mind how savage and bloodthirsty today's *Papio* baboons can sometimes be, how much more so might an extra-large baboon of meat-eating persuasion have been back in Africa's far-distant past? Take, for instance, the very aptly-named *Dinopithecus* ('terrible monkey') *ingens*, which inhabited South Africa during the Pliocene epoch, 5 million to 2.5 million years ago. Adult males were up to 7 ft long, 5 ft high, and 200 lb in weight. This monstrous monkey shared its domain with our distant ancestor *Australopithecus africanus*, upon which it may well have preyed, because in a notable size-reversal, *A. africanus* was no bigger than a present-day baboon whereas *Dinopithecus* was the size of a full-grown present-day human male. How fortunate, then, that *Dinopithecus* is long-extinct – or is it?

One of the most ferocious cryptids on file is the so-called Nandi bear (also known by a number of different native names), which is said to frequent the Nandi forest of Kenya. Many reports have been documented, but the creatures described in them are so diverse in form that when he analysed a wide range of them in his classic book *On the Track of Unknown Animals* (1958), veteran cryptozoologist Dr Bernard Heuvelmans argued that the Nandi bear was in reality a composite cryptid, i.e. 'created' by the erroneous lumping together of reports describing several totally different animal species. He considered that the principal Nandi bear component creatures were large, all-black ratels or honey badgers, abnormally-coloured hyaenas, aardvarks, and, just possibly, a surviving species of chalicothere (a bizarre, claw-footed prehistoric ungulate) plus a surviving species of giant prehistoric baboon.

Of particular relevance to the giant baboon hypothesis is the koddoelo – the Nandi bear representative that allegedly frequents the dense forest of the lower and middle valley of the Tana River, which at just over 600 miles long is the longest river in Kenya. During the early years of the 20th Century, the District Officer for that Tana region was a Mr Cumberbatch, who provided the following details to British anthropologist C.W. Hobley, and which in 1912 Hobley published in the *Journal of the East Africa and Uganda Natural History Society*:

> ...the German missionaries who have lived for many years at Ngao state that the Pokomo natives know of a forest beast called the 'Koddoelo,' and one is said to have been killed near Ngao some years back. On one occasion one of the missionaries found that the whole population of the biggest Pokomo settlement in Kina Kombe district had deserted their village and crossed the river because this animal was roaming about in the bush near the village.

> The animal was described to the District Officer by a Pokomo (who, however, admitted that he himself had not seen it) as being as large as a man, as sometimes going on four legs, sometimes on two, in general appearance like a huge baboon, and very fierce.

Although not common, bipedalism in baboons is by no means unknown. In 1976, for instance, Dr M.D. Rose published a detailed paper in the *American Journal of Physical Anthropology* documenting bipedal behaviour within a troop of olive baboons, noting that it occurred in a wide variety of situations, of which feeding was by far the most common one. Moreover, in a *Folia Primatologica* paper from 2013, Drs F. Druelle and G. Berillon produced quantitative data revealing that in a captive situation, bipedal posture occurred more frequently in juvenile than in adult olive baboons.

In 1913, Hobley published further details regarding the koddoelo in the *Journal of the East Africa and Uganda Natural History Society*, deriving his information from a Mr Rule, who had enquired about it among the Wa-Pokomo people. Here is a detailed description of this cryptid as supplied by the Wa-Pokomo to Rule and thence to Hobley, together with an additional account that had been passed on to him by Tana's then-Assistant District Commissioner:

> Colour, reddish to yellow; length, about 6 feet; height, about 3 feet 6 inches at the withers; hair long, and all accounts agree on the point of a thick mane; tail short and very broad; claws very long; head, fairly long nose, teeth long but not so long as a lion; fore-legs said to be very thick.

> The Pokomo state that several have been killed, and one man says that he killed one himself a good many years ago. It is said to be very fierce, and to visit villages and carry off sheep. On these occasions the natives either cross the river until it leaves the neighbourhood or frighten it away by beating drums. The Waboni hunters know the beast well, but say that they prefer to leave it alone.

> The Assistant District Commissioner on the Tana also sends a further account of the animal, based on recent inquiries, and it was described to him by Pokomo, who said they had seen it, and their account was as follows:- Light in colour, long hair on neck and back, usually goes on fore-legs but can go on its hind-legs, not known to climb trees, rather smaller than a lion, tail about 18 inches long and some 4 inches broad, is nocturnal in its habits, fore-legs very thick; said to leave a track with one deep claw mark behind the marks of its four toes (this is rather obscure). They are agreed about its ferocity, and say it attacks a man on sight. One is said to have killed a rhino near Makere, but this is rather difficult to credit. One tried to raid a goat kraal last January, but was driven away by the noise made by the villagers when the alarm was given.

It is fully confirmed that baboons in South Africa will kill and devour sheep and goats. Rhinos, conversely, are another matter entirely – unless *Dinopithecus* or something like it has indeed persisted into the present day? Of course, the idea of such a large, distinctive, aggressive creature surviving undiscovered by science in modern times seems very remote. Nevertheless, there is little doubt that the above descriptions do recall a baboon in overall morphology, but one of seemingly much bigger size than any species officially existing today.

Engraving from 1833 of a chacma (Cape) baboon *Papio ursinus* holding itself upright upon its hind legs

Having said that, it has long been known that there are a few reports on file concerning exceptionally large specimens of known modern-day baboon species – including none other than *Papio anubis* of Pumphouse Gang fame. One such report was published by the earlier-mentioned Colonel Charles R.S. Pitman more than 70 years ago, in his second of two autobiographical books, *A Game Warden Takes Stock* (1942), and reads as follows:

> Several years ago I was informed that an outsize race of baboons frequented certain parts of the forest [Uganda's Mabira Forest]. They occurred either singly or in small parties, were reputed to be nearly the size of a full-grown man, extremely wary and rarely seen. Further, it was stated that the attitude of these nasty creatures towards unarmed natives

was extremely truculent and, naturally the local folk were very afraid of them. I was promised a specimen if one could be obtained. Eventually an enormous example was collected, but though interesting it proved to be nothing new, the scientific verdict at the British Museum (Natural History) identifying with *Papio anubis anubis* – the green baboon.

Similarly, when considering the Nandi bear in his first autobiographical book, *A Game Warden Among His Charges* (1931), Pitman had stated:

Some of the male baboons I have seen on the Uasin Gishu Plateau [in Kenya] have been of colossal size, capable of killing children with ease; large dogs have been almost torn to pieces, the victim held in its arms. The ape [sic – baboons are of course monkeys] practically disembowels it with downward sweeps of its muscular nail-tipped legs. A great male baboon indistinctly seen in grass or amidst bushland might well be taken for an unknown species.

So if native descriptions of the koddoelo have been liberally seasoned with a generous helping of exaggeration coupled with imperfect observations, it may well be that this belligerent cryptid is simply based upon extra-large specimens of modern-day baboon species, without having to contemplate the formidable prospect of *Dinopithecus* resurrection.

Even so, a giant baboon has also been entertained as a plausible identity for two further Nandi bear incarnations – the chemosit and the kerit. In his own book's coverage of this subject, Heuvelmans postulated that perhaps surviving descendants of *Dinopithecus*:

...have given rise to the widespread legends about the chemosit and koddoelo? At all events a reconstruction of a giant baboon is extraordinarily like most of the natives' and many of the settlers' descriptions of them.

Sadly, however, this entire issue may well remain forever within the realm of unsubstantiated speculation, because Nandi bear sightings seem to have entirely dried up. Indeed, I am unaware of any reports of such encounters from within the past few decades, leading to the inevitable conclusion that even if it were indeed a real creature, the Nandi bear may now be extinct.

Yet perhaps this may not be such a bad thing. After all, as judiciously noted by Heuvelmans:

If such a beast survived it is easy to see why the chemosit arouses such terror. The mere thought that there may be a living animal as huge and strong as a gorilla and as brutally savage as a baboon is frightening enough.

Amen to that!

Chapter 20:
SHOEBILL, HAMMERHEAD, AND BOATBILL - THREE OVERTLY ODD BIRDS

A creature which resembles a very marvel of fairy-land – I mean *Balaeniceps rex*...a name as remarkable as the bird itself, which it has earned from the atmosphere of fable with which it is surrounded, owing to its fantastic form – 'the whale-head' and 'king!' – and verily with him the innermost and obscurest realm of the world is revealed.

Dr Alfred Edmund Brehm – *Bird Life*

People in Africa most often see the bird standing in pools of water, staring intently at its reflection. It is, they say, the one who stands alone, who cannot be pointed at, but who points out wizards and has access to their power. Pursued by wind and rain, this bird is known as the rainmaker, as a herald of the thunderstorm. The people treat the Hamerkop with elaborate respect, keeping their distance, but watching constantly for omens and portents in its behaviour. Their regard is tinged with fear and coloured by the belief that sometimes, perhaps once in many generations, Lightning Bird takes it upon itself to appear among them in human form.

Lyall Watson – *Lightning Bird*

The truth is that we really don't yet know the functional morphology (how it works in an engineering sense) or adaptive significance (why it is advantageous in an evolutionary sense) of the unusual bill that gives the boat-billed heron its name - it is an interesting question for a future graduate student in biology to answer.

Robert Mulvihill –
Pittsburgh Post-Gazette, 8 January 2014

Traditionally, Ciconiiformes is the taxonomic order of birds housing such familiar forms as the herons, egrets, and bitterns; the ibises and spoonbills; and the storks (but see later here for a major update regarding this order's contents). It has also long laid claim to three species so unusual that they cannot be easily allied either with any of the afore-mentioned birds or with one another, and have baffled and bewildered ornithologists ever since they became known to science, as now revealed.

An 1870s engraving depicting shoebills in their natural swampland habitat

SHOEBILL – THE WHALE-HEADED KING

Perhaps the oddest is *Balaeniceps rex*, whose scientific name translates as 'whale-headed king' – but as its head bears little resemblance to a whale, this is a somewhat strange name, even for as strange a bird as *Balaeniceps*. Standing up to 5 ft tall, its general appearance is that of a large, round-shouldered stork with slaty blue-grey plumage and an untidy crest, but distinguishing it instantly from any genuine stork is its enormous, grotesque beak. Roughly 8 in long, with a sharply hooked tip, this incongruous structure greatly resembles a clog-like shoe, earning *Balaeniceps* a much more apt and more commonly used name - shoebill. Similarly, the Arabs call this bird Abu-markub, 'Father of the shoe'.

A peculiar but characteristic behavioural attribute of the shoebill is its tendency to stand perfectly still for lengthy periods of times. Recalling this, a correspondent recently informed me that when she was a child visiting her first zoo, she saw what she initially assumed to be a statue of some strange dinosaur-bird, because it was completely immobile. Fascinated, she stood and looked at it unsuspectingly for a time - until, without warning and in best Talos tradition (fans of the classic Ray Harryhausen film *Jason and the Argonauts* from 1963 will know exactly what I mean here!), this 'statue' slowly turned its head until it was staring directly at her! Its steely gaze peering down its huge beak into her face totally petrified the poor little girl, who was convinced that she was about to be torn apart and devoured!

Happily, the shoebill is in reality a shy, inoffensive species, inhabiting the relatively inaccessible papyrus marshes and floating swamps of the Upper Nile and its central East African tributaries, where it uses its massive beak to catch and extract prey such as fish, water snakes, and frogs (possibly even small mammals, and young crocodiles too) from the surrounding vegetation. It was once believed that its shoe-like shape was a specific adaptation for scooping lungfishes out of the mud, but as lungfishes do not form this species' principal diet, that idea seems unfounded - just as unfounded, it would appear, as many of the assumptions put forward over the years regarding this species' taxonomic affinity to other birds.

Science first became aware of the shoebill in the early 1840s. In his *Expedition to Discover the Sources of the White Nile, in the Years 1840, 1841* (1849), German explorer Ferdinand Werne reported that on 15 December 1840 his party saw a remarkable bird that seemed to them to be as large as a young camel, with a huge pelican-like beak, but lacking the pelican's characteristic pouch. This was undoubtedly a shoebill; sadly, Werne was asleep at the time and his party was unwilling to wake him, so he never observed it himself. Eight years later, however, German ornithologist-explorer Baron Johann W. von Müller was more fortunate, catching sight of two shoebills. Moreover, upon his return to base at Khartoum, Sudan, he saw a pair of dead specimens for sale offered by a slave-dealer. The price that he was asking was too high to interest Baron von Müller, but not long afterwards they were purchased by a traveller from Nottingham named Mansfield Parkyns, who brought them back to England when he returned with various other animal specimens collected during his African sojourn. These were studied by the eminent bird painter John Gould, who prepared a formal scientific description of the shoebill, presented on 14 January 1851 at a meeting of London's Zoological

Shoebills – not singular storks but peculiar pelicans, relatively speaking?

Society and published later that year in its *Proceedings*. And in 1860, Britain's first pair of living shoebills arrived at London Zoo, courtesy of Welsh traveller John Petherick.

Meanwhile, the controversy concerning this species' relationship to other birds had begun in earnest. Gould had classified it as an aberrant, long-legged pelican; but other ornithologists did not agree with that, and tended to ally it either with the herons or with the storks. Today, the shoebill is generally categorised as the sole living occupant of its own family, discrete from both the herons and the storks. The reason for the phenomenal difficulty in satisfactorily classifying this bird rests with its anatomy and behaviour, which embrace a perplexing potpourri of features drawn from at least three different bird families - and two different orders.

Powder-downs are pairs of strange feathers that are never shed, but perpetually fray at their tips to yield a powder that the bird rubs into its other feathers. Herons have three pairs, and the shoebill has a single pair, but storks have none. Also in common with herons, the shoebill's rear toe is held at the same level as its three forward-pointing toes (the rear toe is raised in storks); and when it flies, the shoebill tucks its head and neck backwards, like herons once again.

Even so, whereas in herons the stapes (birds' only middle-ear bone) is primitive in form, avian evolutionist Dr Alan Feduccia showed that it has an identical derived shape in the storks and the shoebill (*Nature*, 21 April 1977). Also agreeing with the storks: the shoebill's middle toe is less than half the length of the tarsus (heel-bone), and it has no webbing between its toes (herons have a partial web between 2-3 of theirs). It displays the storks' beak-clattering behaviour too.

Yet as if all of this were not already sufficient to demonstrate the shoebill's transitional form and conduct, its skull exhibits certain features similar to those of a completely separate order of birds – Pelecaniformes, the pelicans. Furthermore, in true pelican style, it flies with its large beak resting on its breast. Indeed, in 1957 the detailed skeletal studies undertaken by former British Museum ornithologist Dr Patricia A. Cottam on the shoebill convinced her that Gould had been correct all along, that this enigmatic species really was most closely related to the pelicans. However, other researchers (notably Dr Joel Cracraft in the ornithological journal *Auk*, 1985) dismissed its similarities as examples of convergence, i.e. they reasoned that because the shoebill and the pelicans exist in similar habitats and have similar lifestyles, they have evolved into similar forms, even though they originate from separate ancestral stocks.

Then in February 1986 after having directly compared the shoebill's DNA with that of herons, storks, and pelicans, researchers Drs Charles G. Sibley and Jon E. Ahlquist announced in *Scientific American* that, contrary to all expectations, the shoebill's DNA most closely matched that of the pelicans! They published further data in support of their finds during the 1990s. In 2001, extensive research involving DNA hybridisation as well as nuclear and mitochondrial DNA sequence analyses by a team headed by Dr Marcel van Tuinen added further support for the pelicans, the shoebill, and another enigmatic species called the hammerhead (see later here) being of monophyletic (common) origin. Comparable results were also obtained from a comprehensive osteological study by Frankfurt-based ornithologist Dr Gerald Mayr, published in 2003. Modern studies thus offer persuasive evidence for believing, as Gould had proposed over 160 years ago, that the shoebill is basically an aberrant pelican, and that its cranial affinities with pelicans signify direct kinship rather than deceptive evolutionary convergence.

In recent years, the shoebill's classification has also attracted attention because the results of more detailed genetic comparisons (notably the extensive phylogenomic study by Dr Shannon J. Hackett and a large team of co-workers - *Science*, 12 July 2008), involving many different avian genera and families, have required taxonomists to carry out a major overhaul of the contents of the orders Ciconiiformes and Pelecaniformes. These changes can be summarised as follows.

Traditionally, Ciconiiformes has contained the following families: Ardeidae (herons, egrets, bitterns), Balaenicipitidae (shoebill), Scopidae (hammerhead), Ciconiidae (storks), Threskiornithidae (ibises and spoonbills), and Phoenicopteridae (flamingos). Pelecaniformes, meanwhile, has contained Phaethontidae (tropic-birds), Fregatidae (frigate-birds), Sulidae (gannets and boobies), Anhingidae (anhingas or darters), Phalacrocoracidae (cormorants and shags), and Pelecanidae (pelicans). However, modern genetic studies have shown fairly convincingly that Pelecaniformes is polyphyletic, i.e. its families do not all originate from a single common ancestor, but in reality seem to constitute three entirely separate evolutionary lineages.

One of these lineages consists of the tropic-birds, which therefore are now housed within their own separate order, Phaethontiformes. A second lineage comprises the frigate-birds, gannets and boobies, anhingas, and cormorants and shags. So these have all been grouped together within their own new order too, Suliformes. This means that only one original family remains within Pelecaniformes – the pelicans, constituting a third separate lineage.

However, genetic studies have also shown, somewhat unexpectedly, that within the order Ciconiiformes are certain families – namely, the herons, bitterns, and egrets; the shoebill; the hammerhead; and the ibises and spoonbills – that are more closely related to the pelicans than they are to the remaining ciconiiform families. In short, Ciconiiformes is also polyphyletic.

Consequently, these pelican-allied families have now been removed from Ciconiiformes and placed alongside the pelican family within Pelecaniformes. And the flamingos have been allocated their own separate order, Phoenicopteriformes. This means that the only family now remaining in Ciconiiformes is the storks.

Having said that, however, some taxonomists believe that the New World vultures' family, Cathartidae, and the extinct teratorns' family, Teratornithidae, are actually more closely related to the storks than to the Old World vultures or any other birds of prey. Consequently, they have duly included Cathartidae and Teratornithidae alongside the stork family Ciconiidae in Ciconiiformes.

Meanwhile, the shoebill is nowadays back to where it began when first formally described during the 1800s, as a member of the pelican order. However, it is seen, along with the hammerhead, as a taxonomic link between Pelecaniformes and Ciconiiformes.

Even the shoebill's fossil antecedents have stimulated taxonomic turmoil. Until its reclassification in 1980 by Dr Pierce Brodkorb as an ancestral shoebill (based upon the finding of a tarsometatarsus that revealed this taxonomic affinity), *Goliathia andrewsi* had been classed as an aberrant heron. It was first described in 1930 by Hungarian palaeontologist Dr Kálmán Lambrecht, following the discovery of an ulna bone dating from the early Oligocene, which had been obtained in the Jebel Qatrani Formation within Egypt's Fayum Province. The only other widely-accepted fossil relative of the shoebill is *Paludavis richae*, with remains found in Tunisia and Pakistan, but these are more recent, from the Miocene.

Of cryptozoological interest is the shoebill's implication in a very curious case of mistaken identity. From time to time, reports emerge from various remote regions of Central Africa describing alleged sightings of large, macabre-looking creatures soaring through the skies and bearing an impressive resemblance to those long-extinct flying reptilians, the pterosaurs. However, as noted by zoologist Dr Maurice Burton (*Animals*, 18 February 1964) and subsequently explored in greater detail by me within my book *In Search of Prehistoric Survivors* (1995), it is more than likely that many of these involve shoebills. There is no doubt that this strange bird has a distinctly prehistoric appearance, especially when viewed in flight, and anyone unfamiliar with the striking spectacle of its huge, 8.5-ft wingspan and giant beak could certainly be forgiven for thinking that they had spied an aerial anachronism, a cryptic creature supposedly dead for more than 64 million years.

HAMMERHEAD – THE STORM-INVOKING LIGHTNING BIRD

Just under 2 ft long and resembling a short-legged stocky heron with sombre earth-brown plumage, the hammerhead (aka hamerkop, hammerkop, and hammerkopf – all of Afrikaans origin) derives its names from its long, backward-pointing crest which, when carried horizontally, resembles the nail-pulling end of a claw hammer. Moreover, as its crest and its large pointed beak collectively resemble the outline of an anvil, this peculiar bird is also known as the anvil-head.

Widely distributed along the riverbanks, marshes, and ponds of tropical Africa, Madagascar, and Arabia, the hammerhead was first brought to scientific attention during the mid-1700s, in Senegal, by French traveller-naturalist Michel Adanson. In 1760, it was formally described by French zoologist Marthurin J. Brisson, who christened it *Scopus umbretta* - 'with broom and small shade'. Explaining this seemingly odd choice of name, the hammerhead's bushy crest allegedly reminded Brisson of a broom, whereas, somewhat curiously, its crest and beak supposedly reminded him of a small sunshade! Having said that, it has also been suggested that 'umbretta' derives from 'umber', which is another name for the earth-brown shade of this bird's plumage.

Any mystery regarding the hammerhead's name, however, pales into insignificance compared to that which still surrounds its precise relationship to other birds. Just like the shoebill, it embodies an ambiguous assemblage of characters that at the same time link it to and separate it from both the heron family and the stork family.

The hammerhead's heron-like attributes include the incomplete encircling of its bronchi (air tubes) with cartilage - the gaps are sealed with membrane; the pectinate (comb-like) shape of its middle toe's claw; and its rear toe's alignment at the same level as its forward-pointing ones.

1890s engraving of hammerheads with their enormous nests

Yet its lack of powder-downs suggests an affinity with storks, as does the extension of its neck in flight. Electrophoretic examination of its egg-white proteins by Sibley and Ahlquist in 1972 also revealed a correspondence with storks.

Conversely, the hammerhead's general behaviour is neither heron-like nor stork-like. And its parasitic lice (useful indicators of evolutionary affinity between species, as closely related host species often have closely related parasites) are most similar to those of plovers, which belong to an entirely different avian order - Charadriiformes. In the past, some workers had suggested that its nearest relative was the shoebill, but as the latter bird's own classification was still in a state of taxonomic flux, this was not particularly illuminating!

Most recently, however, as noted earlier, and based upon the findings of extensive genetic studies, both the hammerhead's family and that of the shoebill (as well as that of the herons) have been removed entirely from the order Ciconiiformes and rehoused within Pelecaniformes. Consequently, this allies them more closely with the pelicans and the herons than with the storks.

For many years, the hammerhead's fossil ancestry was unknown. In 1984, however, Dr S.L. Olson documented an early Pliocene representative, named *Scopus xenopus*, from Langebaanweg, in South Africa's Cape Province.

The hammerhead is famous behaviourally for boisterously cavorting in wild, highly vocal display dances when associating in small flocks during the breeding season, which varies from one locality to another. Otherwise it is a rather silent, unassuming bird, patrolling the shallow waters of ponds in search of fishes, amphibians (especially the clawed toad *Xenopus*), water insects, and the occasional snail or worm, which it hunts by disturbing the mud at the bottom of the pond with its partially-webbed feet or its slightly-hooked beak.

Curiously, the hammerhead has inspired many strange superstitions and legends. For example, in certain parts of its range it is referred to as the lightning bird, because the local tribes attribute it with the power to invoke terrifying storms at will, and they are also convinced that it can command floods and control the rain. The Kalahari bushmen believe that if anyone tries to rob its nest they will be struck by lightning, and that killing this bird will displease the evil deity Khauna. Another of its titles is 'the King of Birds', because the natives widely believe that other birds help it to build its nest, by bringing it offerings of twigs and leaves.

This odd idea probably stemmed from the enormous size of the hammerhead's nest - which measures up to 4.5 ft in breadth and 6 ft in height, weighs as much as 200 lb, is composed of up to 10,000 sticks, and is sufficiently capacious to house a fully-grown human. It does seem hard to believe at first that this immense edifice could be constructed by two such modest-sized birds as a pair of hammerheads, and yet there is no scientific evidence at all to support the claim that they receive assistance from other species. Ironically, the exact reverse is true. As disclosed in a superb Anglia TV documentary film entitled *The Legend of the Lightning Bird* (first screened in Britain on 20 April 1984), other birds frequently take pieces *away* from the hammerhead's nest, to use in their own! Nevertheless, such blatant theft clearly does not dissuade this species from nest-building – on the contrary, and whether breeding or not, hammerheads construct 3-5 of these huge nests per year.

Small, dark, and sinister is how the hammerhead is unfairly portrayed in many native myths and superstitions

Another local belief, vehemently affirmed by the Xhosa, a Bantu people from South Africa, is that this prodigious nest is divided internally into three distinct 'rooms' - a bedroom for hatching purposes, a dining room for feeding and food storage, and a general hallway. This was also seriously subscribed to by many renowned scientists at one time, including Dr Richard Lydekker (in *The Royal Natural History*, 1894-6). Yet although observations have since confirmed that there are various partitions and ledges inside the nest, there is no evidence for the existence of discrete rooms. Equally, there is no proof that the hammerheads store food inside the nest.

Its nests are so huge that several other animal species often make their homes inside too, including monitor lizards and large snakes, which probably explains folkloric belief in this bird as a shapeshifter. After all, if a hammerhead is seen entering the nest and a big lizard or snake is then seen coming out of it, non-scientific observers steeped in traditional rumour and superstition can be forgiven for drawing an ostensibly evident yet totally erroneous conclusion. Another, more amusing piece of folklore related to its nest is that whenever anyone living in hammerhead territory has their hair cut, they must take great care to collect every last snippet afterwards, because if the hammerhead finds even the smallest tuft and decorates its nest with it, the hair's former owner will assuredly go bald!

This distinctive species also has a widespread reputation among native tribes as a harbinger of doom, presumably because of its somewhat sinister appearance when poised motionless at the side of a pool - a dark, sombre silhouette, with its unique hammerheaded outline. And when staring fixedly into the water in this manner, it is said to be gaining visions of the future. Amazingly, it is a bird of such ill omen that many locals will desert their homes or villages if a hammerhead should as much as fly overhead, as they fear that death will otherwise occur there! Similarly, should one of these birds be heard calling during the evening, and especially if it calls three times in succession, someone will supposedly die during the night. And in Madagascar, natives believe that anyone who destroys its nest will contract leprosy. Moreover, if a hammerhead should fly towards white-water rafts on the Zambezi, the rafting guides will frantically wave their arms, scream, and shout as loudly as possible in order to scare it away, because they firmly believe that bad luck will ensue if it should fly over the rafts.

Such notoriety is totally undeserved, as the hammerhead is a thoroughly harmless, inoffensive species – normally. However, Nos. 124 and 126 of the *Witwatersrand Bird Club News* contain reports of hammerheads aggressively seeing off various birds of prey! Nevertheless, it is no bad thing for it to be burdened with such a reputation, for it actually operates in the bird's favour. This is because natives consider it highly unlucky to hurt or kill a hammerhead, so the species enjoys a protected existence, exempt from the depredations of humankind.

BOATBILL – ITS BEAK IS A TRULY SENSITIVE SUBJECT

Present throughout Central America and northern South America, and named by Linnaeus in 1766, the boatbill *Cochlearius cochlearius* is reminiscent of a night-heron, with its black crown and crest, pale silvery-grey body plumage, and an overall length of roughly 2 ft. Yet unlike night-herons (and all other herons), the boatbill has four pairs of powder-downs instead of only three, and nine primary wing feathers rather than ten. Most distinctive of all, however, is its highly unusual beak. Measuring 3 in long and 2 in wide, this proportionately large, broad structure resembles a sturdy scooping instrument, totally unlike the slimmer beaks of other herons. Indeed, this species' binomial name derives from 'coclearum' - Latin for 'spoon'. Superficially, its beak recalls the shoebill's, but whereas the latter bird's is concave on top, sloping inwards, the boatbill's is convex on top, rising upwards, rather like an upturned boat - hence this bird's English name.

Its specialised, highly modified form implies that the boatbill's beak must have a very precise function, but this remains a controversial subject. One might expect such a beak to be used as a

19ᵗʰ-Century engraving of boatbills

1860s engraving showing the boatbill's superficial similarity to a night-heron

scoop, ladling sluggish, inactive prey from the ample supplies of mud in the mangrove swamps that are the boatbill's favoured habitat. Yet there are reports of boatbills feeding upon highly mobile creatures, such as fishes, frogs, and even mice. How could it catch such fast-moving animals with so cumbersome a beak? Faced with these paradoxes, some researchers have suggested that the boatbill's extraordinary beak has nothing to do with its feeding habits, having evolved instead for display purposes during the breeding season, as both sexes do include beak-clattering behaviour during breeding displays. However, most workers support a feeding role for it too.

Observations of boatbill activity in the wild have verified that it is largely nocturnal – also indicated morphologically, by virtue of its unusually large eyes - and that the sense of touch may well be a crucial factor in its hunting prowess. Certainly, a notably large, broad beak would be much more efficient than a short, slender one in bringing the bird into contact with aquatic prey, and research has since confirmed that its beak is indeed extremely sensitive, capable of detecting even the faintest movements in the water where its prey lives.

Consequently, this species' unique beak has probably evolved primarily to maximise its prey-capturing success, so that this night-active bird does not have to rely solely upon its observational and hearing skills. Instead, it can also achieve notable success by using its remarkable beak to detect its prey's movements in water, and even to make physical contact with it.

Due to its exceptional beak, as well as the differences from other herons noted earlier, the boatbill has been allocated its own family by some researchers, and has even been allied with the shoebill. Nevertheless, the presence of powder-downs (albeit of an atypical number), a pectinate claw on the middle toe of each foot, and an overall anatomy corresponding most closely to the basic heron condition, have satisfied the majority of ornithologists that it is similar enough to warrant inclusion somewhere within the heron family – but where?

Certain workers treat it as a highly specialised offshoot from the night-herons. Conversely, Dr Frederick H. Sheldon's DNA hybridisation studies have indicated that it is most nearly related to the tiger bitterns (*Tigrisoma* spp), constituting with *Tigrisoma* a separate group from all other species included in the heron family.

A heron, therefore, it may well be. Yet the boatbill is evidently no less contentious than are the shoebill and the hammerhead - truly, three overtly odd birds!

Chapter 21:
THE SHAMIR AND THE STONE WORMS

> There are ten things that were created on the eve of Sabbath, and these are they: The
> mouth of the earth (that swallowed Korach); the mouth of the well; the mouth of the
> ass; the rainbow; the manna; the staff; the shamir; the Written Torah; the writing;
> and the Tablets.
>
> *Mishnah*, Avos 5:6

There are a number of mysterious and controversial biblical beasts with potential relevance to cryptozoology, of which the most celebrated examples are undoubtedly Leviathan and Behemoth. Much less famous but no less remarkable than those two cryptids, however, is the small yet highly intriguing subject of this present chapter - the shamir.

THE ENIGMATIC SHAMIR – A ROCK-SLICING LASER GAZER?
Also spelled 'samir' or 'schamir', 'shamir' is the Hebrew name given to a tiny worm-like creature referred to in certain Jewish holy books, including the Midrashim and the Talmud (particularly the Gemara – the component of the Talmud that consists of rabbinical analysis of, and commentary upon, an earlier work known as the Mishnah).

According to Jewish tradition contained within these and other sources, the shamir was one of ten miraculous items created by God at twilight upon the Sixth Day of the Hexameron (the six days of Creation). Although it was only the size of a single grain of barley corn, the shamir was so incredibly powerful that its merest gaze was sufficient to cut through any material with ease, even through diamond itself, the hardest substance on Earth. Such a wondrous creature needed to be safeguarded, so God entrusted the shamir to the hoopoe *Upupa epops* (or to the woodcock *Scolopax rusticola*, or to the moorhen *Gallinula chloropus*, depending upon which version of this legend is consulted), commanding it to protect the shamir from all harm.

In order to contain this mighty if minuscule worm, the hoopoe placed it among a quantity of barley corns, then wrapped them all up together in a woollen cloth, which in turn was placed inside a box

---

The shamir as depicted in the Rosslyn Missal (an Irish manuscript dating from the late 13th or early 14th Century)

fashioned from lead – the only material strong enough to contain the shamir effectively but without disintegrating from the intensity of its laser-like gaze. So here, safely and comfortably ensconced within its leaden domicile, which was retained by the hoopoe in the Garden of Eden, it passed through all the ages that followed.

Only once did the shamir emerge – during the time of Aaron and Moses, when God commanded the hoopoe to lend this worm to Him for the etching of the names of the 12 tribes of Israel upon the precious stones on 12 special priestly breastplates (the Hoshen), one breastplate for each of the tribes and each breastplate composed of a different stone. The task was a very difficult one, but when these stones were shown to the shamir, this astonishing creature accomplished it so expertly that not a single atom of precious stone was lost or destroyed.

After this, the shamir was placed back inside its lead casket, entrusted once more to the hoopoe's care, and there it remained, in undisturbed obscurity – until the time of King Solomon the Wise. Solomon wished to erect a glorious temple in Jerusalem, but he was very mindful of God's instructions, laid down long ago to Moses, that no place of worship, not even an altar (let alone a temple), should be constructed using any tool made from iron. This was because iron is a substance of war, and if anything related to war should ever touch a place of worship, it would be instantly and irrevocably defiled. But if Solomon could not use iron tools, how could the stones needed for constructing his temple be hewn?

In an attempt to solve this riddle, Solomon enquired far and wide, and eventually he learnt about the incredible stone-searing shamir. Determined to utilise its extraordinary power, Solomon dispatched a servant to seek out this wonderful creature and bring it back to him. After a long search, the servant succeeded, and Solomon duly employed the shamir to cut the rocks required for building his celebrated temple – the First Temple in Jerusalem. But that is where the story ends abruptly – because after this magnificent edifice was completed, the shamir allegedly lost its power, then vanished, and has never been heard of again...or has it?

In his engrossing book *Sacred Monsters* (2nd edit., 2011), Rabbi Natan Slifkin wondered if the shamir might have been based upon a real but not particularly well known creature native to Israel's Negev Desert. Namely, the so-called rock-eating snail *Euchondrus*, represented there by three closely-related species - *E. albulus*, *E. desertorum*, and *E. ramonensis*. Less than half an inch long, these mini-molluscs eat lichens that grow beneath the surface of rocks, and they use a toothed tongue-like organ known as the radula to rasp away the intervening rock with great ease and rapidity. However, if such snails were indeed the identity of the shamir, surely the holy books and scriptures would have alluded to their very conspicuous whorled shell? Yet no mention exists of the shamir possessing any such structure. Also, these sources state categorically that the shamir does not destroy any portion of the rocks or precious stones that it cuts through, unlike the activity of these snails.

WAS THE SHAMIR RADIOACTIVE?
Intriguingly, there is an alternative school of theological thought postulating that the shamir was not a living creature at all, but was in fact a mineral itself - specifically an exceptionally hard green

stone, which could cut through all other substances. Yet this identification fails to explain how the stones needing to be cut could be by merely being shown to the shamir, i.e. without the shamir making any direct contact with the stones, using only its gaze to achieve its appointed task. As noted by Rabbi Slifkin, however, one maverick scientist proposed an extremely ingenious, and plausible, solution to this dilemma.

Immanuel Velikovsky (1895-1979) is best-remembered for his highly controversial theories of catastrophic global events producing profoundly revised datings of major events in ancient history, as propounded in bestselling books such as *Worlds in Collision* (1950) and *Earth in Upheaval* (1955). Turning his attention to the enigma of the shamir's identity, Velikovsky suggested that perhaps it was a radioactive substance, which could certainly explain some of the more notable riddles relating to it.

For instance: such a substance could indeed produce its effects upon other substances merely by having them placed near (or shown) to it, not requiring direct contact with them. Also, what better container for a radioactive substance to be housed safely inside than a casket of lead, which would very effectively shield the outside world from this substance's potent effects? And as its radioactivity would diminish with time (i.e. its half-life), this could explain why the shamir's potency had ultimately faded away by the time that King Solomon's temple had been completed. If the shamir were truly a living creature, conversely, its abilities could not be explained by any theory of this nature.

VERMES LAPIDUM - THE MEDIEVAL STONE WORMS
In any event, I had always assumed that this incredible entity was entirely mythical – until 28 November 2013, that is, when Facebook friend Robert Schneck very kindly brought to my attention an astonishing but hitherto exceedingly obscure mystery beast that seemed at least on first sight to be a veritable shamir of the Middle Ages. Robert revealed to me two engravings of bizarre-looking beasts known as vermes lapidum or stone worms, and which had appeared in a hefty German tome authored by Eberhard Werner Happel and entitled *Relationes Curiosae, oder Denckwürdigkeiten der Welt*, which was originally published in five volumes between 1683 and 1691.

According to Happel, the stone worms had originally been brought to public attention by a 17th-Century monk called de la Voye, hailing from a Normandy monastery, who in 1666 had written a letter to a Lord Auzout describing his remarkable discovery. One day, de la Voye had found some of these very small, decidedly odd-looking creatures moving about incessantly inside various holes of their own making in an old wall, much of whose rocky composition had allegedly been eaten away and converted into dust by the devouring nature of the worms. When he pulled out some of them and examined them under a magnifying glass, the monk observed that they were each the size of a single barley corn (the very same description, intriguingly, as used in the Jewish holy books for the shamir) and enclosed in a grey shell, as depicted in the first (labelled Fig. 1) of Happel's two engravings presented here.

As quoted by Happel in his book, the monk continued his account of the stone worms in his

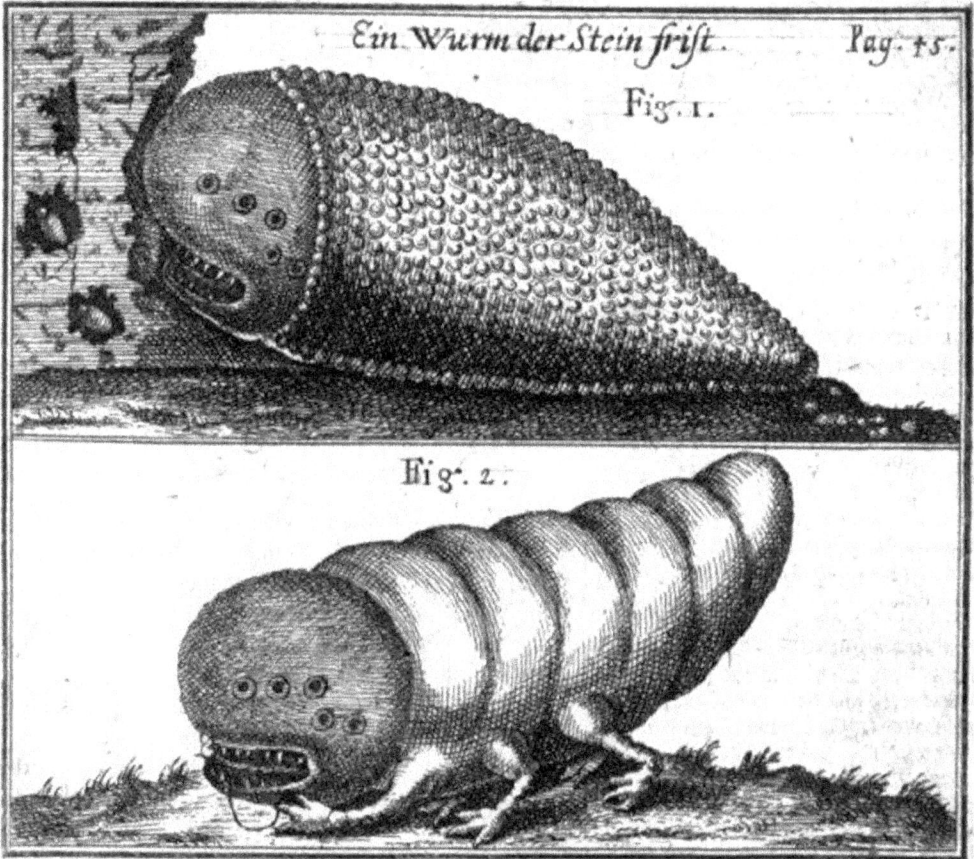

Two engravings (Figs 1 and 2) of alleged stone worms from Eberhard Werner Happel's *Relationes Curiosae, oder Denckwürdigkeiten der Welt* **(1683-91)**

letter to Lord Azout as follows:

> ...on the tip [of the worm's body] there is a hole, through which the excrements can be excreted. On the other end there is a larger hole, trough which the head can be protruded.
>
> They are entirely black, the body shows various segments, near the head there are three legs, each has two joints, not dissimilar to these of a flea.
>
> When they move, their body is suspended in air, the mouth but is still oriented to the rock. The head is bulky, a bit smooth, similar in shape and colour to the shell of a snail...also the mouth is similar large, with four kinds of teeth disposed in cross like manner.

The second engraving (Fig. 2) presumably shows the stone worm in a more advanced state of development than in Fig. 1, as it is now equipped with three pairs of legs. However, both forms seem only to possess small, primitive, laterally-sited ocellus-like eyes (round and black, according to de la Voye), rather than large, compound eyes. This indicates that if the stone worm is an insect, as seems at least remotely possible, it is a larval form rather than an adult (because larval insects do not possess compound eyes, only ocelli).

Conversely, some authors have sought to discount the stone worms as (very) fanciful representations of funnel-weaving spiders, three pairs of legs rather than four notwithstanding and the stone worms' reputed rock-devouring proclivities discounted as apocryphal. Perhaps the presence of multiple ocelli, a characteristic of many spiders (which never possess compound eyes like most adult insects do), influenced their choice of an arachnid identity for these creatures, as there seems little else that would have done so? Certainly, the heavily segmented abdomen of the creature in the second engraving, and the seemingly limbless, shelled form of the creature in the first one, present major problems in reconciling them with any spider.

To be honest, however, the creatures depicted in these two engravings are so bizarre that it is impossible to identify them confidently with any known animal form. If they were indeed real, and not a hoax perpetrated by de la Voye, we can only assume that these engravings are exceedingly fanciful representations, so much so that the worms' true morphology has been enshrouded in artistic exaggeration or error.

As for the worms' stone-devouring behaviour, this too is extremely baffling. Perhaps de la Joye saw these animals amid the wall's crumbling masonry and wrongly presumed that they were directly responsible? Who can say? All that can be stated is that except for brief mentions in a few early 18th-Century dictionaries of natural science, the stone worms rapidly faded into total scientific oblivion shortly after Happel's book was published.

Could it be that, as a monk, de la Voye was well-read across a wide spectrum of religious tracts, was therefore familiar with the mythical shamir from the holy books of Judaism, and had mistakenly thought that the creatures that he had discovered were similar? In reality, however, his stone worms' ostensible comparability to the shamir does not stand up to close scrutiny. For whereas the latter beast disintegrated and annihilated rocks using its formidable, basiliskian gaze, the stone worms actually devoured rocks and stones, at least according to de la Voye's testimony.

Almost 350 years have passed since de la Voye wrote his thought-provoking letter documenting the stone worms, but its subjects remain as mystifying and as unsatisfactorily 'explained' today as they were then - just like the shamir, in fact. Even so, unless the entire episode of their discovery was indeed a hoax and a nonsense, the stone worms must have been something – but what?

Chapter 22:
CONFESSIONS OF A CRYPTOZOOLOGIST - THE SERENDIPITY SCENARIO

The occurrence and development of events by chance in a happy or beneficial way.

Definition of 'serendipity' – *Oxford English Dictionary*

According to the *Oxford English Dictionary*, the word 'serendipity', originating in 1754 when coined by Horace Walpole, was suggested by a Persian-derived Venetian fairy tale, published in 1557 as *Peregrinaggio di Tre Giovani Figliuoli del re di Serendippo,* and later translated into English as *The Three Princes of Serendip* (Serendip being the Persian name for Sri Lanka). In this story, the titular trio of heroes "were always making discoveries, by accidents and sagacity, of things they were not in quest of". More than 250 years later, serendipity certainly plays its part in modern-day cryptozoological research and discovery too, as I have learnt at first hand many times over the years, and have already revealed earlier in this book. So now, to conclude it, here are a few more examples.

THE BIRTHDAY PRESENT THAT CHANGED MY LIFE
It is a very long time ago now, but I still fondly recall a birthday present bought for me by my mother, Mary D. Shuker, that was quite unassuming in form but which changed my life forever, literally...

As a zoologist, media consultant, and author specialising in cryptozoology, it is supremely ironic that my introduction to this enthralling subject was anything but promising.

It all began one day during the early 1970s when, aged around 13, I walked into the department store Boots in the town of Walsall in the West Midlands, England, and spotted a copy of the Paladin abridged paperback edition (1972 reprint) of Dr Bernard Heuvelmans's classic cryptozoology book *On the Track of Unknown Animals*. Picking it up, I opened it at random, and my eyes alighted upon the following sentence, on p. 199, concerning a South

My much-read, greatly-treasured original copy of the Paladin
1972 abridged paperback reprint of *On the Track of Unknown
Animals* (Paladin/Dr Karl Shuker)

American subterranean cryptid called the minhocão:

> Senhor Lebino also related that in the same district, a Negro woman,
> who was going one morning to draw water, found the pool destroyed
> and saw an animal 'as big as a house' crawling away on the ground.

"An animal as big as a house! How ridiculous!", I thought to myself, and promptly replaced the offending book on the shelf.

That hasty action could well have ended my cryptozoological career before it had even begun, cruelly stifled at birth. Happily, however, Fate decreed otherwise. My mother, who was with me, had seen that I'd been reading this particular book, albeit briefly, so she later bought a copy as a surprise birthday present for me in December of that same year. When I opened the parcel and saw which book was inside, I remembered what I had read before, and viewed it with suspicion, but as soon as I began reading it properly, from the beginning, I was of course totally hooked. In no time, I had read it from cover to cover on so many occasions that I could recite great chunks of it.

Eager to learn more about undiscovered animals, I began collecting every newspaper cutting, magazine article, and book that I could find on the subject. And when, while studying zoology at university, I became friends with a fellow student who actually owned the unabridged 1958 hardback edition of *On the Track...*, containing several short chapters that had been omitted from my abridged paperback version, and who kindly permitted me to take his book to the nearest photocopier, I felt as euphoric as if I had been handed the Holy Grail!

How I wished then that I could make a career out of cryptozoology, but, like most zoology students starting university, I anticipated pursuing a traditional professional life as a research scientist.

And indeed, I went on to obtain both a BSC (Honours) degree in pure zoology and also a PhD in zoology and comparative physiology. However, I had always enjoyed writing, and during my spare time I had by now also amassed a formidable private archive of cryptozoological material. So instead of continuing with mainstream zoological research following my PhD, I decided instead to plunge into the uncertain waters of freelance journalism - specialising in cryptozoology and other areas of so-called 'fringe' science.

Some of my early cryptozoological friends and colleagues outside Britain jokingly said that I seemed to spring up from nowhere, and I can appreciate that this may well have been how it looked to them. However, I can assure everyone that, in the time-honoured showbusiness tradition with such matters, it took me quite a long time to become an overnight success!

It began with a number of articles and letters of mine published in various British newspapers, and regular cryptozoological commissions from a British ma-gazine (sadly long-defunct now) called *The Unknown*. Following this, I experienced my first success abroad, with a selection of articles accepted for publication in *Fate* by its then editor, Jerome Clark. My long-running

BERNARD HEUVELMANS

On the Track of
Unknown Animals

With an Introduction by

GERALD DURRELL

My copy of the original 1958 hardback English edition, which I finally
purchased more than a decade after receiving the paperback edition from
Mom as a birthday present (Rupert Hart-Davis/Dr Karl Shuker)

'Alien Zoo' column in *Fortean Times* began soon afterwards (and still appears today), as well as regular articles in many other magazines too.

In 1987, I was amazed and ecstatic when my very first book synopsis, proposing an international survey of feline cryptids, was swiftly accepted by a well-respected London publishing firm, Robert Hale. Even before the ink was dry on its contract, I had already begun writing the book, which was published in June 1989 as *Mystery Cats of the World* - achieving great success, staying in print for several years, and bringing my name to widespread attention at last. Two years later saw the publication of a second book of mine by Robert Hale - *Extraordinary Animals Worldwide*, reviewing a diverse range of animal anomalies.

I had also been researching and preparing another book for quite a long time (although initially only as a hobby), on a subject that had never been covered in book form before - new and rediscovered animals of the 20th Century. In 1993, however, five years of exhaustive work on this pioneering project climaxed with its publication by HarperCollins, as The *Lost Ark* - which I successively updated in subsequent years to yield two fully-revised, much-

My copy of the original two-volume French edition of *On the Track of Unknown Animals*, entitled *Sur la Piste des Bêtes Ignorées* and published in 1955 (Librairie Plon/Dr Karl Shuker)

Mom at Crystal Palace, London, in 2010, standing in front of a
statue of a ground sloth – one of many putative cryptids
examined in *OTTOUA* (Dr Karl Shuker)

expanded sequels – *The New Zoo* (2002), and *The Encyclopaedia of New and Rediscovered Animals* (2012).

More than 20 years have somehow raced by almost unseen and unrealised by me since the publication of *The Lost Ark*, and the current tally of books written by me stands at 21, plus another 12 for which I have acted as consultant and/or contributor, as well as countless articles and blog posts, and my continuing editorship of the *Journal of Cryptozoology* (see pp.   for a complete listing of my books and those for which I have acted as consultant/contributor).

And they all owe their existence to that unassuming birthday present bought for me by Mom all those years ago. Today, it is battered, tattered, stapled, and sellotaped together to within an inch of its life, almost read into oblivion by me over the years. Indeed, I eventually purchased a second, near-pristine copy for my cryptozoological bookshelves in my study. But my original copy remains one of my most treasured possessions, safely housed within a bookcase in my bedroom alongside other much-loved books from my childhood and teenage years.

Thank you, Mom, for being my rock, the foundation upon which I have built my entire life, and, by giving me this humble little book as a birthday present over 40 years ago, propelling me along a hitherto-unsuspected but inspirational path - opening my eyes to the fascinating world of mystery animals that led to my lifelong career in cryptozoological writing and research, and which in turn has given me so much happiness and engaged my continuing interest for such a very long time.

How I wish that you were still here, to wish me a happy birthday in this year and in however many more years await me in the future. But although you can no longer do so, I still have my greatly-prized copy of the Paladin *OTTOUA* edition that you bought me for that long-bygone birthday in my youth, and which reminds me with great joy and thankfulness of when you *were* here, of all the many happy birthdays that we did spend together down through the years, and how very lucky and truly blessed my life has been with you in it as my mother.

God bless you, little Mom.

BETWEEN A ROC AND A HARD PLACE

The main shopping centre of Birmingham, England's second largest city, is replete with unique shops and stalls selling all manner of interesting and often quite esoteric items. Even so, the last thing that I expected to find there when browsing in a certain large fancy goods shop about 10 years ago was a sheaf of roc feathers! But that is precisely what I did find, much to my astonishment and delight. Allow me to explain.

According to Eastern legend and lore, the mighty roc or rukh - featuring as one of Sinbad the Sailor's most formidable protagonists in *The Arabian Nights* - was said to be a monstrous bird of such prodigious size and strength that it could haul elephants aloft, and carry them away to its gargantuan nest where its brood of hungry super-sized offspring would feed upon these hapless pachyderms. For decades, zoologists and cryptozoologists alike attempted to explain

The author with a replica great elephant bird and a life-sized great elephant bird silhouette (Dr Karl Shuker)

Early children's book illustration of Sinbad's ship being attacked by rocs

the mythical roc as having been based upon sightings of a gigantic ostrich-like ratite known scientifically as *Aepyornis maximus*, the great elephant bird, standing 10 ft tall and weighing up to half a ton.

Formerly native to Madagascar, this avian goliath inhabited the extensive marshes and swamps once present on this large island mini-continent, and was known locally as the vouronpatra or vorompatra. It is believed to have survived until at least as recently as the late 1700s, before a lethal combination of over-hunting, introduced avian diseases, and deforestation leading to the drying out of its swampland habitat brought about its demise, but fragments from its enormous eggs can still be commonly found on beaches here.

Unfortunately, reconciling the roc with the great elephant bird faced one major problem. *Aepyornis* was flightless, and therefore incontrovertibly incapable of abducting unwary elephants and swooping off into the air with them, gripped tightly in merciless talons of steel as those dusty Arabian legends would have us believe. Yet due to its huge size, the great elephant bird remained the closest match – indeed, the only remotely plausible match – for the fabled roc...until 1994, that is.

This was when palaeontologist Dr Stephen M. Goodman published a paper documenting the subfossil remains of a hitherto-unknown species of huge eagle, which he formally christened *Stephanoaetus mahery*, the Madagascan crowned eagle, and which is believed to have survived on the island until around 1500 AD (both it and its prey were probably hunted into extinction by humans). This spectacular raptor is thought to have preyed not only upon various now-extinct species of giant lemur (subfossil remains show that some weighed up to 26.5 lb) but also quite possibly upon the great elephant bird itself. Clearly, sightings of this mega-eagle made by early European explorers visiting Madagascar and subsequent exaggeration of those sightings during retellings when back home provide a much more likely explanation for the origin of the roc legends than does the flightless elephant bird.

Yet not even Madagascar's colossal crowned eagle can explain the truly immense plumes that crusaders sometimes purchased in the Middle East to delight and bewilder their families and friends back home in Europe. Claimed by their Arabian vendors to be genuine roc feathers, they sometimes measured 3 ft or more in length, and their vanes' blades were razor-sharp to the touch.

In reality, of course, these spectacular objects were not feathers at all. They were actually the extremely long leaves of the raffia palm tree *Raphia regalis* (and related species). Yet they were

Early children's book illustration of Sinbad being carried aloft by a roc

certainly convincing enough in their superficial resemblance to gigantic feather to fool the unsuspecting crusaders into spending their hard-earned money on them as exotic souvenirs.

Moreover, when I first came upon the tall vase containing a sheaf of these wonderful pseudo-plumes in that Birmingham shop, for just a few moments I too shared the shock and wonder that those crusaders must have experienced when first they saw them. A few stood nearly half as tall as I am (5'10"), and whereas some were brown, others were an exotic jungle green. Once I'd recovered from my surprise, I knew at once what they were, but I still marvelled at finding such cryptozoological curios in such a relatively mundane locality as a high street shop in Birmingham, rather than in some mysterious, shadowy souk within the depths of an Arabian kasbah.

Picking a few up – and soon discovering how painful it was when their sharp vanes stabbed into my hands! – I could definitely understand how their wily vendors in those far-distant Middle

Eastern lands and times had talked the naïve crusaders into believing that these fantastic objects were roc feathers. Indeed, part of me even wanted to believe it myself!

Instead, however, I satisfied myself with the knowledge that here were some wonderful additions to my collection of zoological esoterica, but I received a final shock when I took half a dozen of the green versions (which looked much more feather-like than the plainer brown ones) to the till to pay for them. Incredibly, all six together totalled less than £2, making them without a doubt the best cryptozoological bargains that I had ever purchased!

A few years ago, I found myself back inside that very same Birmingham shop, but its remaining roc feathers were long gone - which, I suppose, is no more than one should ever expect when dealing with imaginary monsters!

One of my prized 'roc feathers' - in reality, like all such plumes, a giant leaf from the raffia palm tree (Dr Karl Shuker)

WHEN OGOPOGO WAS GOING FOR A SONG!

It's not every day that, totally by chance, you encounter a veritable legend, but that's exactly what happened to me one Sunday afternoon during the early 1990s while wandering around a book fair held in the community centre of Kinver, a small Staffordshire village. Looking up at one particular stall, I spotted something that was almost as fabled a cryptozoological artefact as the elusive thunderbird photo itself!

Attached to the side of one of this stall's bookshelves, sealed in cellophane, and on sale for the minuscule sum of just £2, was none other than the original sheet music, complete with fully-illustrated front cover, for 'The Ogo-Pogo - The Funny Fox-Trot'. As every self-respecting cryptozoological enthusiast will confirm, this is the very same English music-hall song from 1924 that gave its name to the now-famous water monster of Canada's Lake Okanagan.

Yet until I saw - and very swiftly purchased! - this cryptozoologically priceless item, it had never been depicted or even accurately quoted from in any crypto-book or article. For upon reading through it, I soon discovered that the lyrics describing Ogopogo's alleged parents ("his mother was an earwig, his father was a whale") were quite different from those various versions purporting to be from it that had been cited in previous works (some of which had replaced 'whale' with 'snail' or had replaced the entire line concerning putting salt on his tail with 'A little bit of head, and hardly any tail'). Here, therefore, are the full, *original* lyrics (by Cumberland Clark, written to music composed by Mark Strong), as given in this sheet music:

> One fine day in Hindustan,
> I met a funny little man.
> With googly eyes and lantern jaws,
> A new silk hat and some old plus fours.
> When I said to that quaint old chap:-
> "Why do you carry that big steel trap,
> That butterfly net and that rusty gun?"
> He replied "Listen here my son:-
>
> I'm looking for the Ogo-pogo,
> The funny little Ogo-pogo.
> His mother was an earwig, his father was a whale,
> I'm going to put a little bit of salt on his tail.
> I want to find the Ogo-pogo
> While he's playing on his old banjo.
> The Lord Mayor of London,
> The Lord Mayor of London,
> The Lord Mayor of London wants to put him in the Lord Mayor's show".
>
> On his Banjo night and day
> The Ogo-pogo loves to play,
> He charms the snakes and chimpanzees,
> The big baboons and the bumble bees.
> Lions and tigers begin to roar:-
> "Play us that melody just once more".
> Did I hear the sound of an old banjo?
> Pardon me I shall have to go!
>
> I'm looking for the Ogo-pogo,
> The funny little Ogo-pogo.
> His mother was an earwig, his father was a whale,
> I'm going to put a little bit of salt on his tail.

The front cover of my sheet music for 'The Ogo-Pogo - The Funny Fox-Trot' (Dr Karl Shuker)

I want to find the Ogo-pogo
While he's playing on his old banjo.
The Lord Mayor of London,
The Lord Mayor of London,
The Lord Mayor of London wants to put him in the Lord
Mayor's show".

Moreover, as can be seen here, the cover portrayed a boot-wearing, antenna-sporting, banjo-playing, pixie-like monster from Hindustan - all far removed indeed from Canada's serpentiform cryptid. Nevertheless, it was this very sheet music that had originated one of the most familiar of all modern-day cryptid nicknames (until then, the Lake Okanagan monster had been known only as the naitaka - a traditional Native American name given to it by the local Okanakane nation).

The timing of my purchase was such that I was able to include a b/w photograph of the Ogo-Pogo sheet music's cover in my 1995 book *In Search of Prehistoric Survivors*, which thus became the very first cryptozoological publication ever to include it, and a year later my next book, *The Unexplained*, became the first publication to include a colour photograph of it.

Thanks to veteran Ogopogo researcher Arlene Gaal, I subsequently obtained a copy of an early American recording of the song itself, performed by the Paul Whiteman Orchestra on an old shellac 78 rpm record. (Please note, however, that in this American version the second verse is missing, and various words and lines in the first verse have been changed from the original English lyrics given above, to yield a much more Stateside-sounding song, in which even 'the Lord Mayor's Show' has been replaced by 'a Broadway show'!)

And so now, after finally mastering the art of uploading music tracks to YouTube, I have pleasure in presenting for your delight on YouTube, the Paul Whiteman Orchestra's version of 'The Ogo-Pogo' foxtrot (check it out at: http://www.youtube.com/watch?v=uQE8T6Ip6Ic).

THE PARK OF THE MONSTERS – A RENAISSANCE GARDEN FOR MEDUSA?
Several years ago, while browsing through a series of bookstalls in the indoor market at Bridgnorth, a town in Shropshire, England, I came upon a hardback travelogue book, dating from around the 1950s/early 1960s as far as I can recall, in which the writer described various sights that he had visited during his journeys around Europe (and possibly elsewhere). One chapter particularly interested me, as it described an extraordinary park located somewhere in continental Europe that was filled with huge, grotesque stone statues of monsters and other bizarre entities, but which had long since been abandoned and was now exceedingly overgrown, heightening its nightmarish aspect.

Normally, anything as unusual as like this would have been enough for me to have purchased the book without hesitation, so I still don't understand why I didn't do so on this particular occasion. To make matters worse, I never even took notice of the book's title or author (I can only assume that I must have had other, more pressing matters on my mind that day!). As

The giant statue of a dragon battling three mammalian antagonists at Bomarzo's Park of the Monsters (Silvana Pellegrini Adam)

always happens in such a situation, of course, I later regretted not purchasing the book and resolved to do so when I was next in Bridgnorth (a town that I often visited back in those days), but, as again always happens, when I did return, the book was gone, and the bookstalls' owner did not even remember it.

Not long afterwards, moreover, he moved out of the market altogether, presumably selling his stock or taking it to set up elsewhere. So any chance of painstakingly going through all of his many books there in case it had been mis-shelved was gone too.

I recently recalled to mind that long-vanished book, so I decided to see if I could discover any information online that may identify the mysterious garden of monsters that it had documented - and, happily, I succeeded! So here is what I found out:

One of Europe's strangest gardens has been dubbed the world's first theme park by some, and is Renaissance by date, but in appearance and content it is decidedly gothic. For although its official name is the Garden of Bomarzo (situated in Viterbo, northern Lazio, in Italy), it is most commonly referred to – and for very good reason – as the Park of the Monsters.

It was created during the 16th Century by Duke Pier Francesco 'Vicino' Orsini (1523-1583), an ex-military officer who was also a leading patron of the arts, and devoted to his wife, Giulia Farnese. When she died, he established the garden (which he called his Sacred Grove) in tribute to her. It consisted originally of a wooded park at the bottom of a deep valley overlooked by Orsini's castle, but he then commissioned the sculpting of a host of arcane gargantuan statues to populate it, some hewn directly from the valley's natural bedrock, and many representing terrifying monsters or other figures from classical Greek mythology. More than two dozen were completed, plus various smaller exhibits such as a temple.

These awe-inspiring but very macabre stone colossi included Cerberus the three-headed hound of Hades, two mermaid-like sirens, a Pegasus fountain, a scarcely-attired reclining nymph, the principal sea god Neptune/Poseidon, the goddess Aphrodite, a giant (possibly Heracles) sculpted in the act of ripping apart another giant (Cacus?), the shape-shifting marine deity Proteus, and a winged woman sitting upon a vast tortoise.

Also present was the bizarre Mouth of Hell – the screaming face of a hideous ogre, whose mouth was a grotto big enough for people to walk through, and inscribed with the words "All reason departs".

Assorted animals included a bear, a whale, and a war elephant of the Carthaginian general Hannibal (carrying a trampled Roman soldier in its great trunk). Perhaps the most outstanding example of Orsini's bizarre statuary, however, was a stupendous sculpture of a winged classical dragon, crouching at bay with jaws open wide as it battled a dog (symbolising spring), a lion (summer), and a wolf (winter).

Placed in an apparently random manner within the park, these daunting goliaths astonished all who saw them, but some felt that their grotesque, melancholic forms and erratic distribution mirrored Orsini's anguished, deranged state of mind resulting from his wife's death. There is controversy as to who designed and prepared the statues. Certain experts attribute them to the celebrated architect Pirro Ligorio. Others support claims that it was none other than Michelangelo who designed these great works, with some of his most talented students sculpting them. A third school of thought suggests that a team of prisoners of war awarded to the duke was responsible.

Following his death, this nightmarish park was gradually abandoned, falling into an eerie state of disrepair, until by the 1800s many of the figures had become virtually hidden within a veritable jungle of overgrown vegetation and unchecked foliage. Orsini's forgotten menagerie of immense megaliths remained neglected and unvisited by all but vagrants and ne'er-do-wells (plus Salvador Dali in 1938, a visit that inspired his 1946 painting 'The Temptation of Saint Anthony') until as recently as 1970. This was when a successful restoration was initiated by the Bettini family owning the land containing this most surreal of gardens.

Today, the Park of the Monsters is a major tourist attraction. Countless visitors wander now through its shadowy realm to encounter its frozen fauna of horror, where Orsini's dragon, winged steeds, triple-headed hell hound, and all of his other weird figures stand forever in stony silence, as if the very gorgon Medusa had cast her evil gaze of petrification upon their grim gathering.

WOLVERINE, WOLVERINE, WHERE HAVE YOU BEEN?

It's always a great feeling when the lost is found, when a mystery is solved – and especially so when the lost had been lost, and the mystery concerning it unresolved, for over 40 years.

Right from a very early age, I had always been fascinated by mysterious and mythological creatures, making my eventual cryptozoological career little short of inevitable. And so it was that a certain episode of a Western show that my family and I viewed one Sunday afternoon on British television during the mid-1960s, when I was around six years old, engaged my attention to a far greater degree than might otherwise have been expected, given the fact that, normally, the Western genre held little if any interest for me.

The episode in question concerned the stalking of a family living alone in the American wilds a century or so ago by a mystifying but greatly-feared beast of such rapacious, belligerent, yet elusive nature that it was referred to superstitiously by the local Native American people as a devil (a name that, as far as I could recall, featured in the title of this episode too). They also called it by another, more exotic-sounding name, and, as in all the best suspense movies, the creature itself was never seen, until the very end. Instead, the viewer had to be content with savage growls, rustles in the undergrowth, and off-screen activity.

Finally, the 'devil' was lured close enough to be shot, and at the denouement it was revealed to be an unexpectedly large specimen of a creature not generally encountered in those parts. But what exactly was that creature? Only at the end was its English name finally given, and it proved to be a species that, as a young child, I had never heard of before – the wolverine.

19th-Century engraving of a wolverine and cubs

Known scientifically as *Gulo gulo*, native to northern North America and also to northern Europe and northernmost Asia, the wolverine or glutton is the largest living terrestrial member of the mustelid family, growing to the size of a small bear. Moreover, it is infamously ferocious, powerful, intelligent, and tenacious, making it one of the most feared species of mammal throughout its range. Little wonder, then, that in the episode it had been referred to by the Native Americans as a devil.

This particular programme had a strong impact on me, as I had been enthralled waiting to discover what the mystery beast in it was, and I can remember watching it avidly a second time a couple of years or so later when the entire series was repeated on television (even though I now knew the creature's identity beforehand). But after that, nothing. As far as I am aware, neither the series in full nor this particular episode from it was ever broadcast on British TV again. But what was the series?

The years passed by, yet despite remembering the wolverine episode in great detail, I never could recall the name of the series itself, and whenever memories of the episode periodically came to mind I always promised myself that I'd pursue this intriguing little mystery, but somehow I never did. Eventually, even the wolverine episode faded in my recollection until it became little more than a hazy, half-forgotten dream. And although I would often flick through books on vintage television, I never obtained any clues as to its series' identity.

Just like its subject, however, the wolverine episode was nothing if not tenacious, and recently it came to mind yet again – but now, armed with the vast research power of the internet, I decided that the time had come to track down this cryptozoological tele-phantom once and for all. I began my search on YouTube, in the hope that the episode, or at least an excerpt or two from it, had been posted there. I knew that I could still remember enough details to be able to recognise it, should it be there. But despite using a variety of key words – 'Western', 'television', 'wolverine', 'devil', '1960s' – nothing promising came up.

So I turned my attention to Google, and used the same key words in its search engine. After a time, I thought I'd discovered it, but it was a false lead. Google had turned up an episode from 1963 called 'The Wolverine' in a Canadian series entitled *The Forest Rangers*, in which a ferocious wolverine turns up in Indian River (the fictional location where this series was set), killing all the livestock there. This plotline certainly compared closely with the programme that I had seen, and the creature was even referred to in it by the same exotic name that I now recalled from 'my' episode – carcajou, an Ojibway name. But when I researched *The Forest Rangers* series, I discovered that its rangers-focused storylines didn't accord at all with those of the series that I had viewed all those years ago. Exit *The Forest Rangers*.

Happily, however, my continued Googling did finally achieve the long-awaited success that I had been hoping for, because there, at last, on my computer screen, was the answer. The series, first broadcast in 1966, was entitled *The Monroes*, which was produced by Qualis in association with 20th Century Fox Television, and included Michael Anderson Jr and Barbara Hershey (playing the parent-substitute figures of big brother and sister) among its stars. It centred around the story of five orphans (aged from 18 down to 6 years of age) trying to survive in 1876 as a family on the frontier in the area around what is now Grand Teton National Park near Jackson, Wyoming, after their father and mother had drowned.

Running for just a single season, *The Monroes*, consisted of 26 episodes – the fourth of which, entitled 'The Forest Devil', was the wolverine episode that had played such a key role in kindling my interest in cryptozoology as a youngster and had afterwards teased and tantalised my mind for over four decades.

And as if my solving this longstanding mystery from my childhood were not satisfying enough, I then discovered that the entire episode had been uploaded on YouTube, in five parts.

Needless to say, and for the first time since the mid-1960s, I duly sat back and watched 'The Forest Devil' – an experience made even more memorable this time by being able to view it in colour. For just like so many other families in Britain at that time, we'd only owned a black-and-white television during the 1960s, so until now I'd never seen this programme in colour.

Returning when an adult to a television show, a book, or even a location that had been so appealing as a child does not always live up to expectations, with the memory of it sometimes proving to have been much more special than the reality. However, I'm happy to report that in the case of 'The Forest Devil', it was every bit as thrilling and enjoyable now, even in these jaded 2010s, as it had been for me back in the 1960s, when everything was still bright and fresh and new and exciting.

In terms of cryptozoological significance, rediscovering 'The Forest Devil' hardly compares, for instance, with my recent solving of the long-baffling Trunko case, nor would it rank alongside the refinding of the legendary thunderbird photograph (should this ever happen one day), but for me, it was just as eventful and satisfying. It also showed me that miracles, even if they are only very minor, personal ones, do indeed still happen in this mundane old world of ours. And that is something else well worth treasuring.

If you'd like to view 'The Forest Devil' on YouTube, here it is in five parts:

Part 1:
http://www.youtube.com/watch?v=s9q4qjulRiA

Part 2:
http://www.youtube.com/watch?v=DTQXskenhDs&feature=relmfu

Part 3:
http://www.youtube.com/watch?v=W6gBXmIzhHM&feature=relmfu

Part 4:
http://www.youtube.com/watch?v=NX7wZGlIjNQ&feature=relmfu

Part 5:
http://www.youtube.com/watch?v=fLagLw5FK9c&feature=relmfu

THE TWO-HEADED KESTREL THAT CAME HOME WITH THE GROCERIES

I may be a cryptozoologist and animal anomalist, but even I have to admit that it's not every day I go into town to buy some groceries and return home with a two-headed kestrel – but 28 September 2012 was one such day!

Browsing in a local market in Wolverhampton, West Midlands, that contains a number of antique/collectors' stalls, I came upon one stall that I hadn't seen before. And there, directly before me, was this truly extraordinary exhibit – a two-headed taxiderm specimen of the

With my newly-acquired two-headed kestrel (Dr Karl Shuker)

With my two-headed taxiderm kestrel (Dr Karl Shuker)

Did this bird get ahead by having two heads? (Karl Shuker)

European kestrel *Falco tinnunculus*.

To cut an extremely short story even shorter: reader, I purchased it! It is an adult female specimen (judging from its brown heads), is in excellent condition; and although I have seen various dicephalous chickens and ducks in the past, this is certainly the very first bicephalic bird of prey that I have ever encountered.

But is it genuine, a bona fide teratological raptor, or – to use an apt falconry term - has it been created to hoodwink its observers? That is a very good question!

What do you think?

Well, let's put it this way. When I encountered this eyecatching taxiderm specimen, I was very surprised to *see* it, but not at all surprised *by* it. And if that in itself sounds surprising – and possibly even a little confusing! – please allow me to explain, via the following sequel, which in reality is a prequel.

Freak shows and travelling circuses usually display at least one two-headed lamb or pig amidst their panorama of curiosities and caprices, so such entities are far from uncommon. The same can certainly not be said, however, for the erstwhile prize exhibit of a private natural history collection amassed some years ago by a West Midlands friend of mine whom I shall simply refer to as Nigel. Among his array of fossils and stuffed animals was a winged wonder

Nigel's photo (top) and one of mine (bottom) (Nigel and Dr Karl Shuker)

Nigel's photograph of his two-headed kestrel (Nigel)

unlike no other – dating from Victorian times, it was a perfectly preserved taxiderm kestrel, with two heads.

Silent testimony to the art of the expert taxidermist, it astonished and entranced all who viewed it, until eventually, along with most of his other specimens, Nigel sold it several years ago.

Only then did he confess the truth to me – it was a fraud, albeit a truly amazing one. The left head was from another specimen entirely, which the taxidermist had meticulously attached to an already-prepared stuffed kestrel from Victorian times, to create this twin-headed falcon.

However, as seen in this photo that he had snapped of his dicephalous marvel some time before selling it (and which, amusingly, depicts the kestrel wearing a gold medallion and chain around its necks!), the artifice was so superb that even though Nigel knew which head was fake, he had great difficulty in detecting the zone of attachment, very deftly concealed beneath its neck feathers. Indeed, to quote the famous words spoken by British television comedian Eric Morecambe concerning his comedy partner Ernie Wise's alleged wig: "You can't see the join!".

After Nigel sold his two-headed kestrel, however, the whereabouts of this wonderful specimen were no longer known, and there seemed little hope of tracing it again. Until that fateful day of 28 September 2012, however, when I was extremely shocked, but delighted, to discover it for sale on a stall in the very same town where Nigel had sold it all those years previously! In any case, I recognised it instantly from Nigel's photograph (a copy of which he had given to me some time ago) – its pose, tree branch mount, and base were all identical, as can be readily confirmed here by comparing Nigel's photo of it with one of mine, snapped on 28 September.

Clearly, therefore, the kestrel had been sold either directly to the stall holder by Nigel, or via one or more intermediary purchasers/owners if such existed. Whatever the explanation, I certainly had no intention of allowing such an astonishing (un)natural history exhibit to disappear into obscurity again – as would almost certainly have happened if someone else had purchased it. So I duly bought it myself.

I hadn't seen Nigel for quite a while, but in February 2014 I bumped into him again, and learnt that he had sold this specimen directly to the Wolverhampton stall holder I'd purchased it from (and who must therefore have been its only owner since Nigel until I came along). Needless to say, Nigel was delighted to know that his unique kestrel has found a good home again! Result!!

HOW I LOST A LEOPON, AND FOUND IT AGAIN!

I previously documented this amazing incident in my book *Cats of Magic, Mythology, and Mystery* (2012), but as a personally-experienced example of serendipity it is so extraordinary that it truly has no equal. Consequently, it definitely bears repeating here.

It's often been said that there is no such thing in life as coincidence, and some of my own experiences certainly support this claim. This next example is a case in point.

As I mentioned earlier in this present book regarding the reverse mermaid (see Chapter 9), even as a child I'd snip articles and reports on strange animals from newspapers and then paste them into a series of large scrapbooks, which constituted the humble foundation of what in later years would become my cryptozoological archives.

When I was around 10 years old, I well remember seeing in one newspaper a very large two-page spread about a truly extraordinary big cat known as a leopon, which proved to be a male hybrid of a leopard and a lioness that had been born in a Japanese zoo. The spread was dominated by a wonderful photograph of this astonishing animal, revealing that its sturdy body was leonine in form but was extensively patterned all over with leopard-like rosettes, and its leopardine head was encircled with a thick lion-like mane. Clearly, it was indeed a crossbreed of these two great cat species, and as I previously hadn't even suspected that they could successfully hybridise and produce viable offspring, let alone had seen photos of any, I was completely captivated by this article and in particular by its amazing photograph.

Consequently, even today, over 40 years later, I still cannot comprehend why on earth I didn't retain the article and paste it without delay into one of my scrapbooks. Yet for some completely unknown reason, I didn't. Instead, rather like a dream in dawn's first light, it somehow slipped away from me and was gone, lost forever from my life, it seemed, but never forgotten.

It was now 1988, and I was earnestly preparing the manuscript of my very first book, *Mystery Cats of the World*, for publication the following year. It included a section on leopons in relation to my coverage of an African crypto-cat known as the marozi or spotted lion, which some investigators have speculated (albeit very implausibly) may constitute naturally-occurring lion x leopard hybrids. Back in those pre-internet days, information on such esoteric creatures was very difficult to trace, and on more than one occasion while writing my leopon section I cursed myself for not having retained that informative newspaper article from my childhood. Then again, I mused, perhaps it might be possible to obtain a copy of it after all. Thinking back, I reflected that the article must have been published in or around 1969 or 1970, and of the wide range of newspapers that

Male leopon at Hanshin Park Zoo, Japan (Dr Warren D. Thomas)

my family read at that time, the most likely one in which it would have appeared, based upon its layout and style as I recalled, would have been London's *Daily Mirror*.

Armed with these admittedly sparse details, I wrote to the archive department of the *Mirror*, and asked if there was any chance that they could trace the leopon article and sell a photocopy of it to me for research purposes in relation to my book. I certainly didn't have high expectations of obtaining any result, but a few weeks later a large flat envelope arrived, and there inside was an A3-sized photocopy of the leopon article! It was dated 28 January 1970, occupied pages 12 and 13, and was entitled 'The lion who just can't change his spots'.

Gazing at it again after almost 30 years, I was thoroughly amazed, but delighted. The photo, by globetrotting *Mirror* cameraman Kent Gavin, was exactly as I'd pictured it in my mind's eye, and there was also a second, much smaller one, showing the leopon's leopard father, Kaneo, and his lioness mother, Sonoko, together. Reading the article's text revealed that the leopon himself – the extremely exotic-looking subject of the main photo - was called Johnny, who lived with his parents and four siblings at a zoo in Kobe. Needless to say, I lost no time in writing a very grateful letter of thanks to the *Mirror* archivist who'd successfully uncovered the much-prized article for me, and I also made sure that this time it was placed safely on file – after which I naturally assumed that this was the end of the lengthy saga of the discarded leopon article.

In fact, I couldn't have been more wrong, because by far the most extraordinary episode was still to occur.

Eleven years had now passed since my mystery cats book had been published to great acclaim in June 1989, and, as is so often the case with me on sunny Saturday afternoons in the spring and summertime, I found myself on that eventful day in question at Hay On Wye. Situated on the English-Welsh border, this is the so-called 'Town of Books', famous worldwide for housing within its modest-sized borders around 30 different second-hand bookshops.

One of my favourites was the Natural History Bookshop (sadly, it closed in 2012), which contained a veritable treasure trove of wildlife books on every conceivable field of natural – and unnatural – history. Outside, there was often a bookcase containing bargains and sell-off volumes, often priced at no more than a pound or so each, which were generally arranged very haphazardly, due to the considerable number of browsers who sifted through them in search of a find.

For reasons that I simply cannot even begin to understand, despite being confronted by shelf after shelf of enticing titles my gaze was instead drawn to a large creamy-white nondescript-looking volume sticking out a little near the bottom of a pile of books on a shelf close to the ground on the right-hand side of this bookcase. I couldn't see its title, and, indeed, could scarcely even see the book itself, so why should I have given it a second glance, let alone felt compelled to pull it out from under the weight of books piled on top of it? And yet, inexplicably, I did.

And once I was holding it, I noticed that there was something sandwiched between two of its pages, something that looked like a folded sheet from a newspaper. Taking it out, I carefully unfolded it, and then heard a mighty clang – the sound of my jaw crashing to the floor, as my eyes struggled to accept what they were looking at.

There, in my visibly trembling hands, were pages 12 and 13 from the *Daily Mirror* for 28 January 1970 – 'The lion who just can't change his spots'!! The pages were a little brown – they were, after all, just over 30 years old – but I scarcely even noticed that. All I could see was Johnny the Japanese leopon's photo staring back at me!

Apparently, there is a one in 14 million probability of winning the jackpot in the UK National Lottery. But what is the probability of my opening a book selected seemingly at random in a town over 80 miles away from my home and packed with countless thousands of other books, only to find inside it a copy of the exact same newspaper article that I had thrown away 30 years earlier and had wished with all my heart ever since that I hadn't? It must be so minuscule as to be virtually incalculable, and yet, somehow, incredibly, I'd done it!

I hardly need state that I immediately went inside the shop, paid my pound for the book, and came out again, still quaking in shock, with its precious newspaper article tucked back safely inside its pages. And as soon as I arrived back home, I took the article out, placed it inside a

The lost is found – the original cutting of the well-remembered leopon newspaper article from 1970 that I serendipitously rediscovered tucked inside a book in Hay on Wye 30 years later! (*Daily Mirror*/Dr Karl Shuker)

transparent plastic A4 packet, and filed it away in the appropriate place within my archives, where it remains to this day - except when, every so often, I take it out again, gently unfold its slightly brittle pages, and gaze with a mixture of awe and rapt reverence at its fondly-remembered photograph. Ironically, I can't even remember what the book was in which I'd found this newspaper article, or even what happened to it afterwards – I suspect that it was given away to a charity shop with various other surplus books of mine when I had one of my periodic weeding-out sessions.

Some writers and readers fervently believe in the existence of a library angel - a supernatural force or entity that helps them to locate a certain book (or other publication) if their need to find it is sufficiently compelling. So was I guided by divine intervention to locate a copy of my long-lost leopon article? If that seems unlikely, how much likelier is it that I could have found it in the way that I did do simply by random chance?

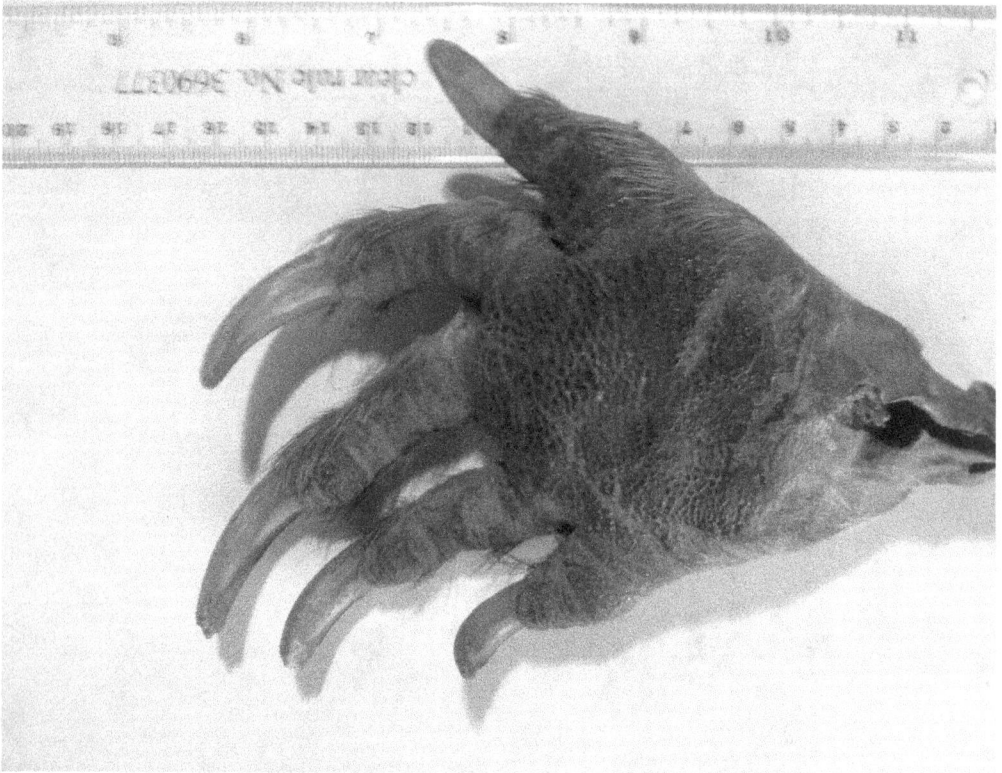

Ben Coult's mystifying preserved paw, palm surface uppermost (Ben Coult)

THE CURIOUS CASE OF THE WEREWOLF PAW...THAT WASN'T!

One thing about being a cryptozoologist is that you regularly receive the most unusual emails, and the following one was no exception! It was sent to me on 5 September 2012 by Ben Coult from County Durham in northern England:

> I have an interesting artefact that I found about 15 years ago and was wondering if you would be interested in me sending you through a few photos of it?
>
> Firstly let me assure you that I am fairly well versed in the subject, having read Heuvelmans and Sanderson for years, and was recently referred to your website by a friend. Also I have a solid upbringing in local natural history and many contacts in that field.
>
> I shall now elaborate, firstly I am from a small village in the north east of England in county Durham (an area not known for unusually large mammals), when I was about 16 myself and a friend were riding bikes

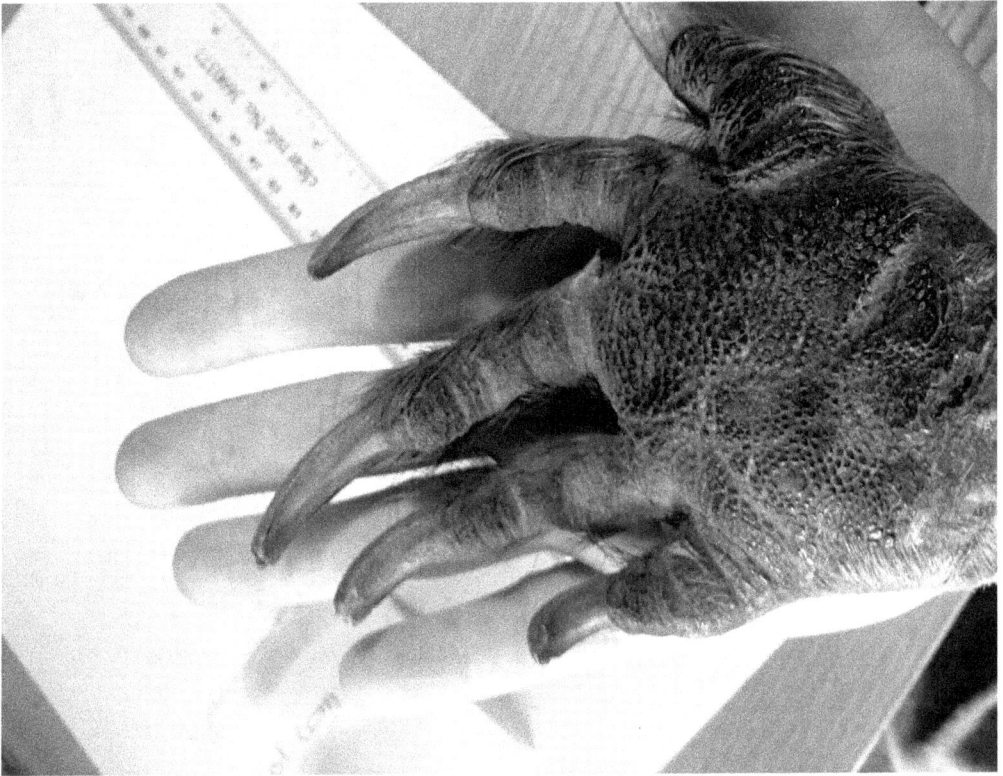

Comparing the paw's size with Ben's own hand (Ben Coult)

through the village when I spotted something in the road which looked for all the world like a poorly stuffed hand. I jumped off the bike and retrieved it and it turned out to be just that. At the age of 16 the 'hand' was only marginally smaller than my own, it has a tough leathery pad on the palm and was obviously heavily haired on the back although (it is obviously quite old) the hair is now sparse. The hand has five 'fingers' each equipped with a 20 to 30 mm [approx. 1 in] claw, these are heavy and curved and look similar to Sloth or Ratel claws.

I have had numerous people look at it including a family friend who is the ex-taxidermist for the Hancock museum in Newcastle and no-one has been able to identify it yet apart from saying 'it could be' or 'it looks like'. Another strange twist to this is that the wrist of the hand has been twisted together and dried, this looks like it must have been done when the specimen was originally stuffed, and a purple ribbon was tied around the pinched in wrist.

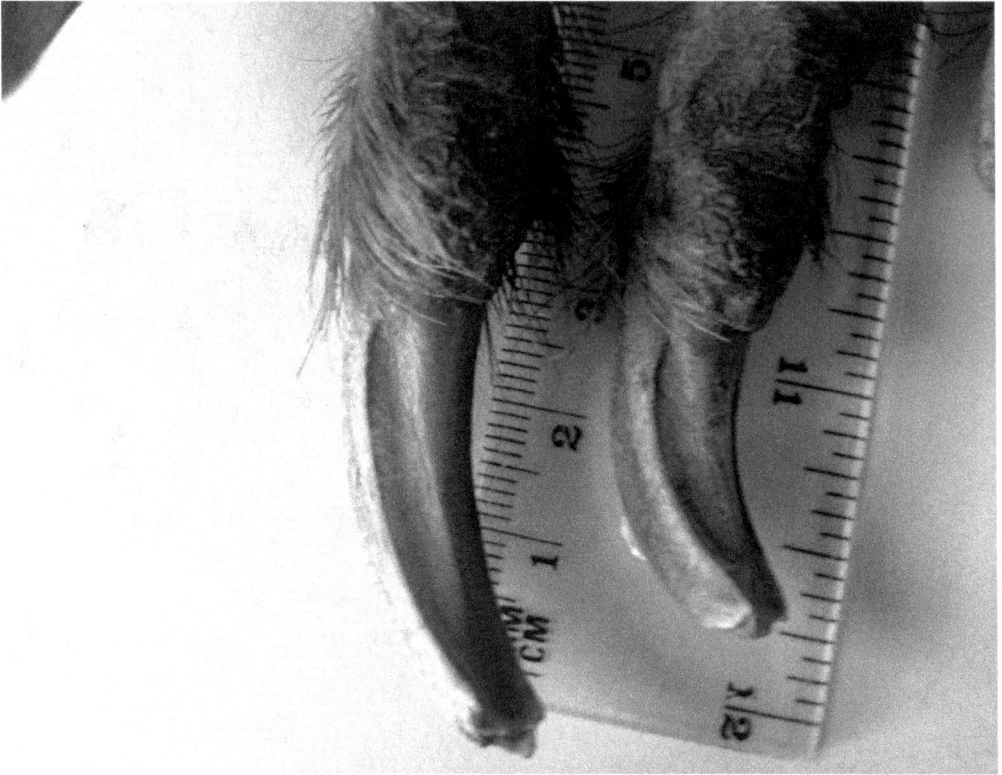

The claws of the paw are very pronounced (Ben Coult)

The hand has been sat on my bookcase for 15 years now although I still try and get people to identify it from time to time, hence this email. If you are interested please drop me an email and I will send through a couple of pictures.

Needless to say, I was extremely interested, as my curiosity had definitely been piqued by this singular incident. So I emailed Ben back straight away with a request to see his photos of the hand, and a few hours later I received five close-up pictures of it, reproduced here with Ben's kind permission.

My initial thought when I looked at them was 'badger', and when a few days later I posted the photo of a badger paw (palm surface uppermost) on my Facebook Wall and on the pages of the various Facebook cryptozoology groups that I've founded, most of the FB friends commenting upon it shared my thought. Yet when I compared Ben's photos of his preserved paw specimen to photos of badger paws, the correspondence did not seem very close at all.

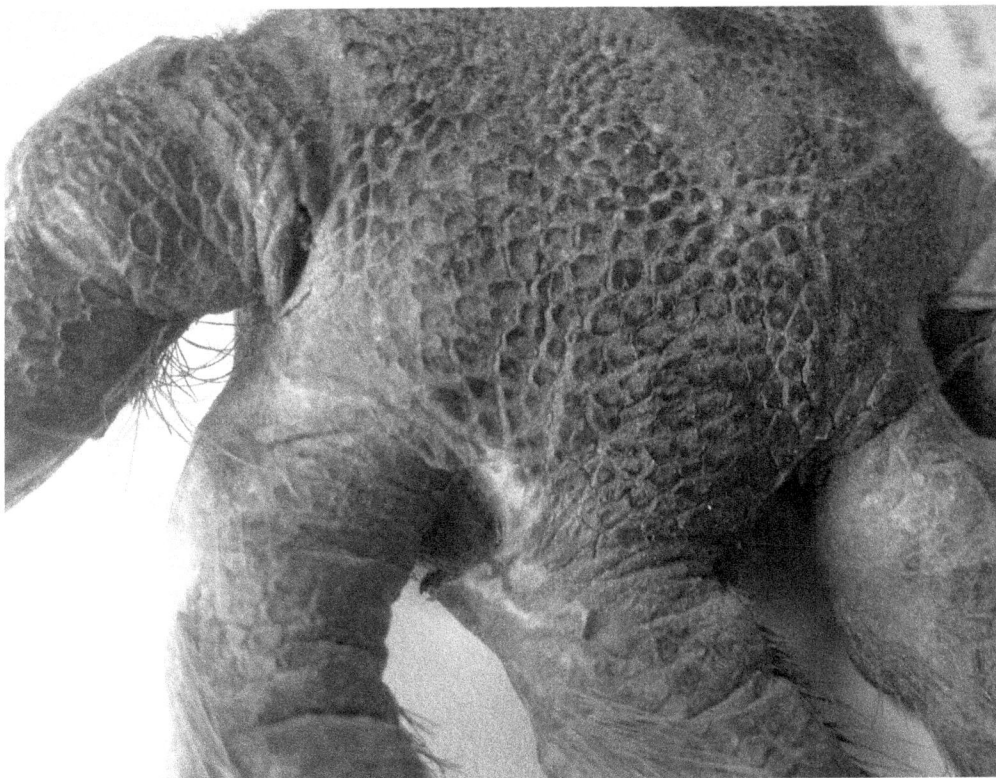

Close-up of the mosaic-like tessellated patterning of the paw's skin (Ben Coult)

In particular, the plantar (palm) pad of the badger was very different in form from that of this paw, and the digits of the badger paw seemed much shorter and broader than those of the latter. There was also the latter's distinctive mosaic-like skin patterning to explain.

Also, if it were indeed nothing more unusual than a badger's paw, why hadn't the professional, highly-experienced taxidermist to whom Ben had shown it been able to identify it?

Perhaps, therefore, as Ben had speculated, it was indeed from something rather more exotic, but what?

Other suggestions posted to my FB pages included bear, monkey, sloth, mole, turtle, and (albeit decidedly tongue-in-cheek!) chupacabra and werewolf!

Someone who shares my own interest in unusual preserved animal specimens is Southampton University palaeontologist Dr Darren Naish. So on 8 September I posted on his Facebook wall this same paw photo and the paw's background details, then awaited his opinion as to its

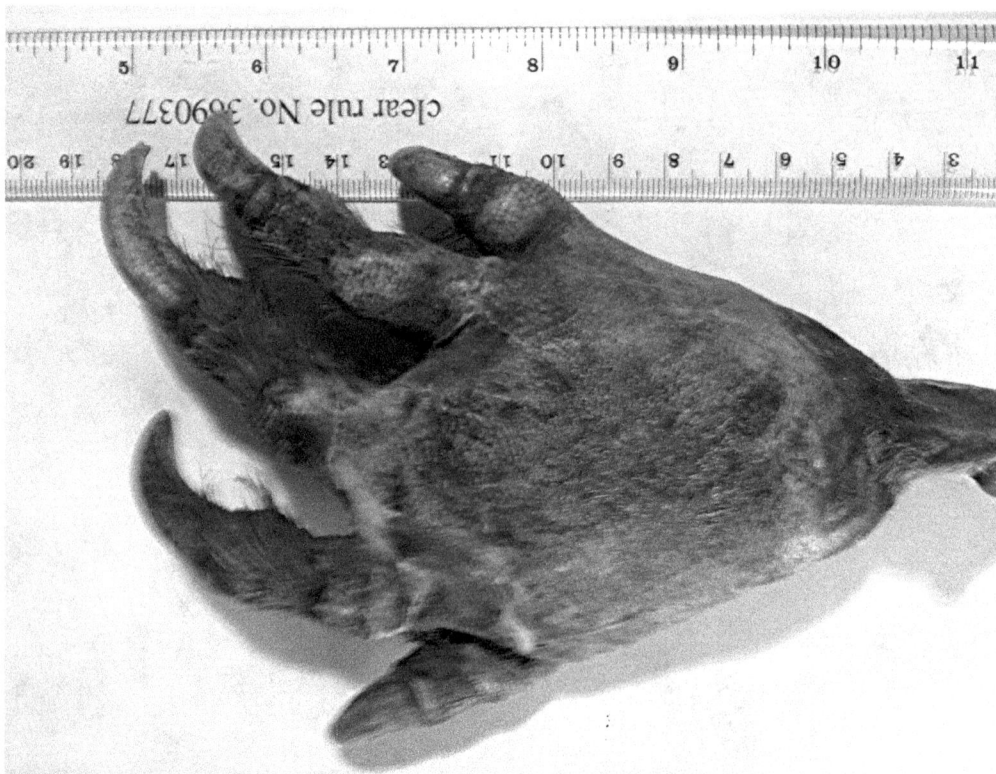

Ben's preserved paw, dorsal surface (Ben Coult)

original owner's likely identity. When I received it later that evening, I was certainly surprised, because Darren suggested that the paw was from a kangaroo!

When he followed up his pronouncement, however, by providing me with the following link - http://3amkickoff.files.wordpress.com/2008/08/kangaroo-hands.jpg   - to an online photograph of a kangaroo's paws, there could be no doubt as to the accuracy of his identification. For the paws in this photo provided a perfect match with Ben's enigmatic specimen, right down to the mosaic skin patterning. German cryptozoologist Markus Bühler also sent me some online photos of kangaroo paws, and I located some as well, thereby providing plenty of independent confirmation for this identity.

Judging from the paw's size, the precise species involved is most likely to be either the red kangaroo *Macropus rufus* (which is the world's largest living kangaroo species as well as the largest surviving marsupial of any species) or the great grey kangaroo *M. giganteus* (the second-largest kangaroo species alive today). Adult males of the red kangaroo stand

My mother Mary D. Shuker and I with a red kangaroo interloper – note its lengthy claws (Dr Karl Shuker)

up to 6.9 ft tall (up to 6.6 ft in the great grey) and weigh around 200 lb or more (around 145 lb in the great grey). Both species are extremely common and used in many commercial ventures.

Nevertheless, one aspect of the mystery of Ben's preserved paw remained unresolved. Why had its wrist been constricted with a ligature of purple ribbon, and why was this paw lying discarded on the road?

Facebook friend Martin Cotterill from Coventry has amassed a very impressive collection and knowledge of taxiderm specimens and other natural history exhibits, and he wondered whether the paw could have been derived from a kangaroo-paw backscratcher. I knew that macabre novelty items like this do indeed exist - some dating back to Victorian times and occasionally met with at antique fairs or in bric-a-brac shops, as well as modern-day ones still widely sold in Australian souvenir shops.

A swift check online revealed photographs of several such objects, consisting of a genuine preserved kangaroo paw attached to a long pole. If the paw from one of these backscratchers were pulled off, it would then need to be tied tightly closed at its base, in order to prevent its stuffing from falling out. So that would explain the presence of the purple ribbon ligature around the wrist of Ben's specimen.

In short, Ben's mystery specimen was the paw from a kangaroo-paw backscratcher, which presumably had either been broken or been deliberately pulled apart in order to obtain the paw as a somewhat grotesque curiosity - which had subsequently been discarded, lying on the road until Ben had cycled by and found it there. It also explains why even his knowledgeable taxidermist friend had not recognised what animal it had originated from. After all, I can't imagine that he'd seen too many demised kangaroos in County Durham!

Another mystery beast case solved, via the collective 'brain' of my contacts network and myself, with particular thanks to Dr Darren Naish, Markus Bühler, and Martin Cotterill, but most of all of course to Ben Coult for so kindly bringing this highly intriguing case to my attention and permitting me to document it here.

WHAT'S IN A NAME?
Finally: I have been asked many times whether I specifically planned a career in cryptozoology or whether it was just something that happened. Perhaps the truth is neither of these – perhaps it was destiny, pre-ordained, the hand of fate. Read the following and judge for yourself.

Right from a child, my surname had always mystified me. I did know that it was German, even though my father's family is entirely English in origin as far back as we can trace (which is several generations). So too is my mother's family. What I didn't know was what it meant. What was the English translation of 'Shuker'? Despite perusing numerous books of surname origins as a child and early teenager, I never managed to find any mention of mine – until one

A preternatural Black Dog (Andy Paciorek / http://www.batcow.co.uk/strangelands/)

day during the mid-1980s, when, while idly thumbing through yet another such volume in a Birmingham bookshop, to my great surprise I found it! But that surprise was nothing compared to what I experienced when I discovered what my name actually meant!

According to that book, 'Shuker' derived from 'Schuck' (I had previously read that 'Shuker' was once spelt 'Schucker'), which was apparently a Germanic term for 'monster'! More specifically, it referred to a goblin-like creature of the night, especially one that could acquire the form of a huge black dog – which may help to explain, therefore, the origin of the name 'Black Shuck' for a famous example from eastern England of the Black Dog zooform phenomenon. In other words, I had a cryptozoological surname - or, at the very least, one that pertained directly to unexplained creatures!

Having said that, I later discovered an alternative derivation for my surname – this time from 'Schuker', an early Germanic name of pre-10th Century origin, which was an occupational term for someone who earned their living by sieving corn by shaking. Nevertheless, the very fact that one translation for 'Shuker' involves a direct link to monsters and mystery beasts is nothing if not intriguing, and would remain so even if that were all – i.e. even if there were no other links between such entities and names appertaining to me. But that is not all – far from it!

Guess what my two nicknames were at school? One, due to the presence of several stone ornaments of that nature in my front garden, was Gnome – a mythical mini-humanoid entity. The other, due to my surname not rhyming readily with any familiar word, was a seemingly meaningless nonsense word, at least as far as the young junior-school children who coined it were concerned. However, it would be instantly recognised as something very meaningful by any cryptozoologist or zoomythologist. For the nickname in question was none other than 'pooker', which, with only the slightest change in spelling, becomes 'pooka' - a legendary Irish monster, taking the form of a huge black dog or goblin pony that carries off unwary children and drowns them.

And as if all of this cryptozoological and zoomythological lexilinking were still not intriguing enough, my maternal grandmother's maiden name was Griffin! In other words, a direct name-link with that famous beast of legend that sports the head and wings of an eagle but the torso, limbs, and tail of a lion.

Even my home town, Wednesbury, is named after a Nordic god - Woden or Odin, thus explaining why you can find here a beautiful gleaming metal statue of Sleipnir, Odin's unique eight-legged steed.

A cryptozoologist by choice, or by destiny? Somehow, I don't think that choice ever came into it, do you?

A pooka (Andy Paciorek / http://www.batcow.co.uk/strangelands/)

BIBLIOGRAPHY

-, 'The Lion Who Just Can't Change His Spots' [leopon] (*Daily Mirror*, London, 28 January 1970).

-, 'Some Fishy News From Cairo' ([Unidentified London daily tabloid newspaper], London, 7 April 1973).

-, 'Antarctic "Rosetta Stone" Provides Clues To Early Southern Hemisphere History' (*Antarctic Journal*, vol. 22, March/June 1982), pp. 1-2.

-, 'The Story Of Eve' [Shropshire mammoths] (*Shropshire Magazine*, vol. 39(11), October 1987), pp. 13-15.

-, 'The Iliamna Lake Monster' (*Alaska Magazine*, vol. 54, January 1988), p. 17.

-, 'Monster Sturgeon' [Lake Washington monster] (*Fortean Times*, no. 50, summer 1988), p. 15.

-, '"Terror Bird" Roamed Antarctic' (*Columbus Dispatch*, Ohio, 12 February 1989).

-, 'Mammoths to Get Home of Their Own' (*Daily Mail*, London, 12 March 1991).

-, *Treasures of the Vatican Library: And To Every Beast...* (Millennium: Alexandria, 1994).

-, 'The Brazilian Invisible Fish' (*The Museum of Hoaxes*, http://www.museumofhoaxes.com/hoax/archive/permalink/the_brazilian_invisible_fish 8 November 2007).

Aburrow, Y., *Auguries and Omens: The Magical Lore of Birds* (Capall Bann: Chieveley, 1994).

Agnolin, F., '*Brontornis burmeisteri* Moreno & Mercerat, un Anseriformes (Aves) Gigante del Mioceno Medio de Patagonia, Argentina' (*Revista del Museo Argentino de Ciencias Naturales*, new series, vol. 9, 2007), pp. 15-25.

Allen, G.M., *Extinct and Vanishing Mammals of the Western Hemisphere* (American Committee for International Wild Life Protection: Washington DC, 1942).

Alston, E.R., 'On the Specific Identity of the British Martens' (*Proceedings of the Zoological Society of London*, 3 June 1879), pp. 468-74.

Alvarenga, H.M.F., and Höfling, E., 'Systematic Revision of the Phorusrhacidae (Aves: Ralliformes)' (*Papéis Avulso de Zoologia*, vol. 43(4), 2003), pp. 55-91.

Alvarenga, H.M.F., *et al.*, 'The Youngest Record of Phorusrhacid Birds (Aves, Phorusrhacidae) From the Late Pleistocene of Uruguay' (*Neues Jahrbuch für Geologie und Paläontologie Abhandlungen*, vol. 256, 2010), pp. 229–34.

Angst, D., *et al.*, '"Terror Birds" (Phorusrhacidae) from the Eocene of Europe Imply Trans-Tethys Dispersal' (*PLoS ONE*, vol. 8(11), 2013), e80357.

Angst, D., *et al.*, 'Isotopic and Anatomical Evidence of an Herbivorous Diet in the Early Tertiary Giant Bird *Gastornis*. Implications For the Structure of Paleocene Terrestrial Ecosystems' (*Naturwissenschaften*, vol. 101(4), April 2014), pp. 313-22.

Aplin, K.P., and Helgen, K.M., 'Quaternary Murid Rodents of Timor Part 1: New Material of *Coryphomys buehleri* Schaub, 1937, and Description of a Second Species of the Genus' (*Bulletin of the American Museum of Natural History*, no. 341, 2010), pp. 1-80.

Armstrong, E.A., *The Folklore of Birds* (Collins: London, 1958).

Auffenberg, W., *The Bengal Monitor* (University Presses of Florida: Gainsville, 1994).

Austin Jr, O.L., and Singer, A., *Birds of the World* (Paul Hamlyn: London, 1961).

Baring-Gould, S., *Curious Myths of the Middle Ages* (Longmans, Green: London, 1892).

Barloy, J-J., 'Rumeurs sur des Animaux Mystérieux' (*Communications*, no. 52, 1990), pp. 197-218.

Baskin, J.A., 'The Giant Flightless Bird *Titanis walleri* (Aves: Phorusrhacidae) From the Pleistocene Coastal Plain of South Texas' (*Journal of Vertebrate Paleontology*, vol. 15, 1995), pp. 842-4.

Bell, R., 'Dawn of the Killer Shrews' [talk given by Raymond Bell to the Edinburgh Fortean Society on 10 March 2009; summarised at http://www.edinburghforteansociety.org.uk/Archives/Lavellan.html].

Bertelli, S., *et al.*, 'A New Phorusrhacid (Aves: Cariamae) From the Middle Miocene of Patagonia, Argentina' (*Journal of Vertebrate Paleontology*, vol. 27(2), 2007), pp. 409-19.

Bille, M.A., *Rumors of Existence: Newly Discovered, Supposedly Extinct, and Unconfirmed Inhabitants of the Animal Kingdom* (Hancock House: Blaine, 1995).

Boer, L.E.M. de, and Brink, J.M. van, 'Cytotaxonomy of the Ciconiiformes (Aves) With Karyotypes of Eight Species New to Cytology' (*Cytogenetics and Cell Genetics*, vol. 34, 1982), pp. 19-34.

Brodkorb, P., 'A Giant Flightless Bird From the Pleistocene of Florida' (*Auk*, vol. 80, April 1963), pp. 111-15.

Brodkorb, P., 'A New Fossil Heron (Aves: Ardeidae) From the Omo Basin of Ethiopia, with Remarks on the Position of Some Other Species Assigned to the Ardeidae' *In:* Campbell, K.E. (ed.), *Papers in Avian Paleontology Honoring Hildegarde Howard* (*Natural History Museum of Los Angeles County Contributions in Science*, no. 330, 1980), pp. 87-92.

Brooks, Roy, 'Unusual Lizard Sightings in North Wales' (*Herne Bay Reptile & Amphibian Club Newsletter*, June 1997), p. 5.

Brown, L.H., *et al.* (eds), *The Birds of Africa, Volume 1* (Academic Press: London, 1982).

Browne, P., *The Civil and Natural History of Jamaica* (Privately published: London, 1756).

Burris, D., 'Koalas and Kangaroos' [contains referenced close-up photo of kangaroo paws] (*3amkickoff*, http://3amkickoff.wordpress.com/2008/08/04/koalas-and-kangaroos/ 4 August 2008).

Burton, M., 'Gone For Ever?' [shoebill as living pterosaur] (*Animals*, vol. 3(11), 18 February 1964), p. 308.

Burton, M., and Benson, C.W., 'The Whale-Headed Stork or Shoe-Bill: Legend and Fact' (*Northern Rhodesia Journal*, vol. 4, 1961), pp. 411-26.

Burton, M., and Burton, R. (eds), *Purnell's Encyclopedia of Animal Life*, 6 vols (BPC: London, 1968-70).

Campbell, J.H., 'Edwin James Report on *Bipes* Reconsidered' (*Herpetological Review*, vol. 11, March 1980), pp. 6-7.

Carleton, M.D., and Olson, S.L., 'Amerigo Vespucci and the Rat of Fernando de Noronha: A New Genus and Species of Rodentia (Muridae, Sigmodontinae) From a Volcanic Island off Brazil's Continental Shelf' (*American Museum Novitates*, no. 3256, 4 March 1999), pp. 1–59.

Case, J.A., Woodburne, M.O., and Chaney, D.S., 'A Gigantic Phororhacoid(?) Bird From Antarctica' (*Journal of Paleontology*, vol. 61, November 1987), pp. 1280-4.

'CDarwin', 'Meet An Ancestor: Theropithecus oswaldi' (*The Caveman's Corner*, http://blogs.scienceforums.net/evoanthro/2008/02/09/meet-an-ancestor-theropithecus-oswaldi/ 9 February 2008).

Chandler, R.M., 'The Wing of *Titanis walleri* (Aves: Phorusrhacidae) From the Late Blancan of Florida' (*Bulletin of the Florida Museum of Natural History, Biological Sciences*, vol. 36, 1994), pp. 175–80.

Chorvinsky, M., 'Duck-Eaters From The Deep?' (*Fate*, vol. 42(9), September 1989), pp. 38-9.

Coleman, L., and Huyghe, P., *The Field Guide to Lake Monsters, Sea Serpents, and Other Mystery Denizens of the Deep* (Tarcher/Penguin: New York, 2003).

Coope, G.R., and Lister, A.M., 'Late-Glacial Mammoth Skeletons From Condover, Shropshire, England' (*Nature*, vol. 330, no. 6147, 3 December 1987), pp. 472-4.

Cottam, P., 'The Pelecaniform Characters of the Skeleton of the Shoe-Bill Stork, *Balaeniceps rex*' (*Bulletin of the British Museum (Natural History), Series D (Zoology)*, vol. 5, 1957), pp. 51-72.

Cracraft, J., 'On the Systematic Position of the Boat-Billed Heron' (*Auk*, vol. 84, 1967), pp. 529-33.

Cracraft, J., 'Monophyly and Phylogenetic Relationships of the Pelecaniformes: A Numerical Cladistic Analysis' (*Auk*, vol. 102, 1985), pp. 834-53.

Cruz, P. de la, 'Gorgeous Grotesques' [Bomarzo's Park of the Monsters] (*Garden Design*, http://pauladelacruz.com/bomarzo.pdf November/December 2009), pp. 72, 74-5.

Dale, C.W., 'The Mammalia of Dorsetshire' (*Proceedings of the Dorset Natural History and Antiquarian Field Club*, vol. 24, 1903), pp. 18-33.

Dallet, R., and Wolff, R., *Animals of Africa* (Litor Publishers: Brighton, n.d. [1960s]).

Dorst, J., and Probst, P., *The Colourful World of Birds* (Paul Hamlyn: London, 1963).

Downes, J., *The Smaller Mystery Carnivores of the Westcountry* (CFZ Publications: Exwick, 1996; updated edition, CFZ Press: Bideford, 2006).

Drinnon, D., 'The Water Leaper' (*Still On The Track*, http://forteanzoology.blogspot.co.uk/2009/11/dale-drinnon-water-leaper.html 22 November 2009).

Druelle, F., and Berillon, G., 'Bipedal Behaviour in Olive Baboons: Infants Versus Adults in a Captive Environment' (*Folia Primatologica*, vol. 84(6), 2013), pp. 347-61.

Duncan, P.M. (ed.), *Cassell's Natural History*, 6 vols (Cassell, Petter, and Galpin: London, 1883-9).

Dundee, H.A., 'A Comment On J. Howard Campbell's *Bipes* Report' (*Herpetological Review*, vol. 11, September 1980), pp. 74, 76.

Eberhart, G.M., *Mysterious Creatures: A Guide to Cryptozoology*, 2 vols (ABC-Clio: Santa Barbara, 2002; updated edition, CFZ Press: Bideford, 2013-14).

Edmeades, B., 'Scary Monsters Can Materialize Out of the Darkness' [*Dinopithecus*] (*Megafauna*, http://megafauna.com/the-book/part-iii/scary-monsters-can-materialize-out-the-darkness/ 2006).

'EHA' [Aitken, E.H.], *The Tribes On My Frontier* (W. Thacker: London, 1904).

Farley, G., "Monster Fish' in Lake Washington? It's Not the First Time' (*King 5 News*, http://www.king5.com/news/pets-and-animals/Eight-foot-long-dead-sturgeon-found-floating-in-Lake-Washington-218555201.html 6 August 2013).

Feduccia, A., 'The Whalebill is a Stork' (*Nature*, vol. 266, 21 April 1977), p. 132.

Fennell, J.H., *A Natural History of British and Foreign Quadrupeds* (J. Thomas: London, 1841).

Fenton, A., and Heppell, D., 'The Earth Hound – A Living Banffshire Belief' (*Scottish Studies*, no. 31, 1992-3), pp. 145-6.

Fleming, J., *History of British Animals* (James Duncan: London, 1828).

Gans, C., and Pappenfuss, T., 'There is No Evidence That *Bipes* Occurs in the U.S.' (*Herpetological Review*, vol. 11, September 1980), p. 74.

George, W., *Animals and Maps* (University of California Press: Berkeley, 1969).

Gibbons, W.J., *Missionaries And Monsters* (Coachwhip: Landisville, 2006).

Gooders, J. (ed.), *Birds of the World*, 9 vols (BPC: London, 1969-71).

Goodman, S.M., 'Description of a New Species of Subfossil Eagle from Madagascar: *Stephanoaetus* (Aves: Falconiformes) From the Deposits of Ampasambazimba' (*Proceedings of the Biological Society of Washington*, no. 107, 1994), pp. 421-8.

Gould, J., 'On a New and Most Remarkable Form in Ornithology' [shoebill] (*Proceedings of the Zoological Society of London*, 14 January 1851), pp. 1-2.

Grant, J., 'Notes on an Alleged Species of Poisonous Lizard, &c' (*Calcutta Journal of Natural History*, vol. 1, 1840), pp. 371-89.

Greenway, J.C., *Extinct and Vanishing Birds of the World*, 2nd edition (Dover: New York, 1967).

Greenwell, J.R (ed.), 'Stafford Lake Monster Caught' (*ISC Newsletter*, vol. 4(4), winter 1985), p. 8.

Grzimek, B. (ed.), *Grzimek's Animal Life Encyclopedia*, 13 vols (Van Nostrand: London, 1972-5).

Grzimek, B. (ed.), *Grzimek's Encyclopedia of Mammals*, 5 vols (McGraw-Hill: New York, 1990).

Hackett, S.J., *et al.*, 'A Phylogenomic Study Of Birds Reveals Their Evolutionary History' (*Science*, vol. 320, no. 5884, 2008), pp. 1763–8.

Hancock, J., and Kushlan, J., *The Herons Handbook* (Croom Helm: London, 1984).

Happel, E.W., *Relationes Curiosae, oder Denckwürdigkeiten der Welt*, 5 vols (T. von Wiering: Hamburg, 1683-91).

Hardwicke, T., *Illustrations of Natural History, Volume 2* (Adolphs Richter and Parbury, Allen: London, 1834).

Hartlaub, G., 'On a New Bird From the Isle of Madagascar' [kinkimavo] (*Proceedings of the Zoological Society of London*, 13 May 1862), pp. 152-3.

Heuvelmans, B., *On the Track of Unknown Animals* (Rupert Hart-Davis: London, 1958).

Heuvelmans, B., *On the Track of Unknown Animals*, abridged edition (Paladin: London, reprinted 1972).

Hichens, W., 'African Mystery Beasts' (*Discovery*, vol. 18, December 1937), pp. 369-73.

Hobley, C.W., 'On Some Unidentified Beasts' (*Journal of the East Africa and Uganda Natural History Society*, no. 6, 1912), pp. 48-52.

Hobley, C.W., 'Unidentified Beasts In East Africa' (*Journal of the East Africa and Uganda Natural History Society*, no. 7, 1913), pp. 85-6.

Holland, R., 'The Palé Beech Marten' (*Still On The Track*, http://forteanzoology.blogspot.co.uk/2009/02/guest-blogger-richard-holland-pale.html 23 February 2009).

Hume, J.P., and Walters, M., *Extinct Birds* (T & AD Poyser: London, 2012).

Ingersoll, E., *Birds in Legend, Fable and Folklore* (Longmans, Green: London, 1923).

Jablonski, N.G. (ed.), Theropithecus*: The Rise and Fall of a Primate Genus* (Cambridge University Press: New York, 1993).

James, E., *Account of an Expedition from Pittsburgh to the Rocky Mountains, Performed in the Years 1819 and '20, by Order of the Hon. J.C. Calhoun, Sec'y of War; Under the Command of Major Stephen H. Long, Volume 1* (H.C. Carey and I. Lea: Philadelphia, 1823).

Jenkin, R., *New Zealand Mysteries* (A.H. & A.W. Reed: Wellington, 1970).

Johnson, P., 'Mammoth Find in a Gravel Quarry' (*Shropshire Star*, Telford, 30 September 1986).

Johnson, P., 'Mammoth Jigsaw Puzzle' (*Shropshire Star*, Telford, 8 October 1986).

Kenny, J.S., *Views From the Bridge: A Memoire of the Freshwater Fishes of Trinidad* (St Joseph: Trinidad and Tobago, 1995).

Kipling, J.L., *Beast and Man in India* (Macmillan: London, 1891).

Knight, C., *Pictorial Museum of Animated Nature*, 2 vols (London Publishing: London, 1856-8).

Krishnan, M., *Eye in the Jungle* (Universities Press: Himayatnagar, 2006).

Lambrecht, K., 'Studien über Fossile Riesenvögel' (*Geologica Hungarica, Seria Palaeontologica*, vol. 7, 1930), pp. 1-37.

Lewis, O., 'Barking Up The Wrong Tree' (*Still On The Track*, http://forteanzoology.blogspot.co.uk/2009/03/oll-lewis-barking-up-wrong-tree.html 6 March 2009).

Loyd, L.R.W., *Bird Facts and Fallacies* (Longmans, Green: London, 1924).

Lydekker, R. (ed.), *The Royal Natural History*, 6 vols (Frederick Warne: London, 1894-6).

MacFadden, B.J., *et al.*, 'Revised Age of the Late Neogene Terror Bird (*Titanis*) in North America During the Great American Interchange' (*Geology*, vol. 35(2), 2007), pp. 123–6.

Mareš, J., *Svet Tajemných Zvírat (Littera Bohemica: Prague, 1997)*.

Marshall, L.G., 'The Terror Bird' (*Field Museum of Natural History Bulletin*, vol. 49, 1978), pp. 6-15.

Maslin, T.P., 'An Annotated Check List of the Amphibians and Reptiles of Colorado' (*University of Colorado Studies, Series D (Physical and Biological Sciences*, vol. 6), pp. i-vi, 1-98.

Matthew, W.D., and Granger, W., 'The Skeleton of *Diatryma*, a Gigantic Bird From the Lower Eocene of Wyoming (*Bulletin of the American Museum of Natural History*, vol. 37, 1917), pp. 307-26.

Mayr, G., 'The Phylogenetic Affinities of the Shoebill (*Balaeniceps rex*)' (*Journal für Ornithologie*, vol. 144, 2003), pp. 157-75.

Meijer, H.J.M., and Due, R.A., 'A New Species of Giant Marabou Stork (Aves: Ciconiiformes) From the Pleistocene of Liang Bua, Flores (Indonesia)' (*Zoological Journal of the Linnean Society*, vol. 160, 2010), pp. 707–24.

Michaux, J., *et al.*, 'A ^{14}C Dating of *Canariomys bravoi* (Mammalia, Rodentia), the Extinct Giant Rat From Tenerife (Canary Islands, Spain), and the Recent History of the Endemic Mammals in the Archipelago' (*Vie et Milieu*, vol. 46, 1996), pp. 261–6.

Miller Jr, G.S., 'A Second Collection of Mammals From Caves Near St. Michel, Haiti' [quemi] (*Smithsonian Miscellaneous Collections*, vol. 81(9), no. 3012, 30 March 1929), pp. 1-30.

Minton, S.A., and Minton, M.R., *Venomous Reptiles* (George Allen and Unwin: London, 1971).

Mourer-Chauviré, C., *et al.*, 'A Phororhacoid Bird From the Eocene of Africa' (*Naturwissenschaften*, vol. 98(10), 2011), pp. 815–23.

Muirhead, R., 'A "Winged Toad" in Suffolk in 1662' (*Flying Snake*, vol. 2(1), November 2012), pp. 9-12.

Musser, C.G., and Newcomb, C., 'Malaysian Murids and the Giant Rat of Sumatra' (*Bulletin of the American Museum of Natural History*, vol. 174, 1983), pp. 327-598.

Naish, D., '2007: A Good Year For Terror Birds And Mega-Ducks' (*Tetrapod Zoology*, http://scienceblogs.com/tetrapodzoology/2008/06/10/2007-year-of-terror-birds/ 10 June 2008).

Newman, H., *Indian Peepshow* (G. Bell: London, 1937).

Newton, A., and Gadow, H., *A Dictionary of Birds* (Adam and Charles Black: London, 1896).

Newton, M., *Encyclopedia of Cryptozoology: A Global Guide* (McFarland: Jefferson,

2005).

Norman, J.R., 'A New Blind Catfish From Trinidad, With a List of the Blind Cave-Fishes' (*Annals and Magazine of Natural History*, vol. 18, 1926), pp. 324-31.

Nozedar, A., *The Secret Language of Birds* (HarperElement: London, 2006).

O'Malley, L.S.S., *Bengal, Bihar, Orissa and Sikkim* (Cambridge University Press: Cambridge, 1917).

Olson, S.L., 'Paleornithology of St Helena Island, South Atlantic Ocean' (*Smithsonian Contributions to Paleobiology*, no. 23, 1975), 49 pp.

Olson, S.L., 'A Hamerkop From the Early Pliocene of South Africa (Aves: Scopidae) From Langebaanweg Southwestern Cape Prov-ince' (*Proceedings of the Biological Society of Washington*, vol. 97, 1984), pp. 736–740.

Parfitt, E., 'Fauna of Devon. Mammalia' (*Transactions of the Devonshire Association*, vol. 9, 1877), pp. 306-30.

Pennant, T., *British Zoology* (Benjamin White: London, 1768).

Pennant, T., *History of Quadrupeds* (Benjamin White: London, 1793).

Perrin, A., *East of Suez* (A. Treherne: London, 1901).

Pitman, C.R.S., *A Game Warden Among His Charges* (James Nisbet: London, 1931).

Pitman, C.R.S., *A Game Warden Takes Stock* (James Nisbet: London, 1942).

Reichenbach, H., *Phantom Fame* (Simon and Schuster: New York, 1931).

Rhys, J., *Celtic Folk-Lore, Welsh and Manx*, 2 vols (Clarendon Press: Oxford, 1901).

Richardson, A., and Hunt, L., 'Twitchers Flock To See Rare Bird' [hoopoe in Walsall] (*Express and Star*, Wolverhampton, 9 October 2006).

Romero, A., and Creswell, J.E., 'In Search of the Elusive "Eyeless" Cave Fish of Trinidad, W.I.' (*National Speleological Society News*, vol. 58(10), 2000), pp. 282-3.

Rose, M.D., 'Bipedal Behavior of Olive Baboons (*Papio anubis*) and Its Relevance to an Understanding of the Evolution of Human Bipedalism' (*American Journal of Physical Anthropology*, vol. 44(2), March 1976), pp. 247-61.

Sanderson, I.T., *Caribbean Treasure* (Viking Press: New York, 1939).

Saunders, J.J., 'Britain's Newest Mammoths' (*Nature*, vol. 330, no. 6147, 3 December 1987), p. 419.

Saville, D.B.O., 'Gliding and Flight in the Vertebrates' (*American Zoologist*, vol. 2, 1962), pp. 161-6.

Selwyn, F., et al., *The Shropshire Mammoth* (Leopard Press: Shrewsbury, 1989).

Sheldon, F.H., 'Phylogeny Of Herons Estimated From DNA-DNA Hybridization Data' (*Auk*, vol. 104, January 1987), pp. 97-108.

Shuker, K.P.N., 'The Shropshire Mammoths' (*The Unknown*, no. 26, August 1987), pp. 33-8.

Shuker, K.P.N., *Extraordinary Animals Worldwide* (Robert Hale: London, 1991).

Shuker, K.P.N., *In Search of Prehistoric Survivors: Do Giant 'Extinct' Creatures Still Exist?* (Blandford: London, 1995).

Shuker, K.P.N., *The Unexplained: An Illustrated Guide to the World's Natural and*

Paranormal Mysteries (Carlton: London, 1996).

Shuker, K.P.N., 'Land of the Lizard King' (*Fortean Times*, no. 95, February 1997), pp. 42-3.

Shuker, K.P.N., *From Flying Toads to Snakes With Wings: From the Pages of FATE Magazine* (Llewellyn Publications: St Paul, 1997).

Shuker, K.P.N., *The Beasts That Hide From Man: Seeking The World's Last Undiscovered Animals* (Paraview Press: New York, 2003).

Shuker, K.P.N., *Mysteries of Planet Earth: An Encyclopedia of the Inexplicable* (Carlton: London, 1999).

Shuker, K.P.N., *Extraordinary Animals Revisited: From Singing Dogs To Serpent Kings* (CFZ Press: Bideford, 2007).

Shuker, K.P.N., 'Two Sea Monster Tales' (*Fortean Times*, no. 285, March 2012), pp. 52-3.

Shuker, K.P.N., *Mirabilis: A Carnival of Cryptozoology and Unnatural History* (Anomalist Books: New York, 2013).

Shuker, K.P.N., *Dragons in Zoology, Cryptozoology, and Culture* (Coachwhip Publications: Greenville, 2013).

Sibley, C.G., and Ahlquist, J.E., 'A Comparative Study of the Egg White Proteins of Non-Passerine Birds' (*Bulletin of the Peabody Museum of Natural History*, vol. 39, 1972), pp. 1-276.

Sibley, C.G., and Ahlquist, J.E., 'Reconstructing Bird Phylogeny By Comparing DNA's [*sic*]' (*Scientific American*, vol. 254, February 1986), pp. 68-75.

Slifkin, N., *Sacred Monsters: Mysterious and Mythical Creatures of Scripture, Talmud and Midrash*, 2nd edition (Zoo Torah: Springfield, 2011).

Smith, H.M., and Holland, R.L., 'Still More On *Bipes*' (*Herpetological Review*, vol. 12, March 1981), pp 8-9.

Stack, J.W., 'On the Disappearance of the Larger Kinds of Lizard From North Canterbury' (*Transactions and Proceedings of the New Zealand Institute*, vol. 7, 1874), pp. 295-6.

Sterndale, R.A., *Natural History of the Mammalia of India and Ceylon* (Thacker, Spink: Calcutta, 1884).

Strum, S.C., *Almost Human: A Journey Into the World of Baboons* (Elm Tree: London, 1987).

Stuckwish, D., *Biblical Cryptozoology: Revealed Cryptids of the Bible* (Xlibris: Bloomington, 2009).

Switek, B., 'Terror Birds Ain't What They Used To Be – A Titanis Take-Down' (*Wired*, http://www.wired.com/2011/02/terror-birds-aint-what-they-used-to-be-a-titanis-take-down/ 2 December 2011).

Taylor, E.H., 'Does the Amphisbaenid Genus *Bipes* Occur in the United States?' (*Copeia*, no. 4, 10 December 1938), p. 202.

Temminck, C.J., and Laugier de Chartrouse, M., *Nouveau Recueil de Planches Coloriées d'Oiseaux, pour Servir de Suite et de Complement aux Planches Enluminées de Buffon*, 102 vols (G. Levrault: Paris, 1820-39).

Thorneycroft, G.V., 'African Palm Civet' (*African Wild Life*, vol. 12, 1958), p. 81.

Thunberg, C.P., *Travels in Europe, Africa, and Asia Made Between the Years 1770 and 1779; in Four Volumes,* 2nd edition (F. and C. Rivington: London, 1795).

Tuinen, M. van, *et al.*, 'Convergence and Divergence in the Evolution of Aquatic Birds' (*Proceedings of the Royal Society of London, Series B*, vol. 268), pp. 1345-50.

Tweedie, M., 'Fossil Elephants In Essex' (*Animals*, vol. 5(10), 15 December 1964), pp. 290-1.

Velikovsky, I., 'Shamir' (*Kronos*, vol. 6(1), fall 1980), pp. 48-50.

Wallace, A.R., *The Malay Archipelago* (Macmillan: London, 1869).

Wallis, R., 'Letter to the Editor' [re cenaprugwirion] (*British Herpetological Society Bulletin*, autumn/winter 1987), p. 65.

Watson, L., *Lightning Bird: One Man's Journey Into Africa* (Hodder and Stoughton: London, 1982).

Werne, F., *Expedition to Discover the Sources of the White Nile, in the Years 1840, 1841* (Richard Bentley: London, 1849).

Wilson, D.E., and Reeder, D.M., (eds.), *Mammal Species of the World: A Taxonomic and Geographic Reference*, 3rd edition (Johns Hopkins University Press: Baltimore, 2005).

Wood, J.G., *Illustrated Natural History*, 3 vols (Routledge, Warne, and Routledge: London, 1859-63).

Woodburne, M.O., *et al.*, 'New Fossil Vertebrates From Seymour Island, Antarctic Peninsula' (*Antarctic Journal*, vol. 22, 1987 review issue), pp. 4-5.

Yalden, D.W., *The History of British Mammals* (T & AD Poyser: London, 1999).

INDEX OF ANIMAL NAMES

Dr Karl Shuker - Cryptozoologist

The author, Dr Karl P.N. Shuker (Dr Karl Shuker)

ABOUT THE AUTHOR

Born and still living in the West Midlands, England, Dr Karl P.N. Shuker graduated from the University of Leeds with a Bachelor of Science (Honours) degree in pure zoology, and from the University of Birmingham with a Doctor of Philosophy degree in zoology and comparative physiology. He now works full-time as a freelance zoological consultant to the media, and as a prolific published writer.

Dr Shuker is currently the author of 21 books and hundreds of articles, principally on animal-related subjects, with an especial interest in cryptozoology and animal mythology, on which he is an internationally-recognised authority, but also including a poetry volume. In addition, he has acted as consultant for several major multi-contributor volumes as well as for the world-renowned *Guinness Book of Records/Guinness World Records* (he is currently its Senior Consultant for its Life Sciences section); and he has compiled questions for the BBC's long-running cerebral quiz 'Mastermind'. He is also the editor of the *Journal of Cryptozoology*, the world's only existing peer-reviewed scientific journal devoted to mystery animals.

Dr Shuker has travelled the world in the course of his researches and writings, and has appeared regularly on television and radio. Aside from work, his diverse range of interests include motorbikes, travel, quizzes, poetry, the life and career of James Dean, Sherlockia, collecting masquerade and carnival masks, philately, world mythology, and the history of animation.

He is a Scientific Fellow of the prestigious Zoological Society of London, and a Fellow of the Royal Entomological Society. He is Cryptozoology Consultant to the Centre for Fortean Zoology, and is also a Member of the Society of Authors.

Dr Shuker's personal website can be accessed at http://www.karlshuker.com and his extremely popular mystery animals blog, ShukerNature, can be accessed at http://www.karlshuker.blogspot.com

His poetry blog can be accessed at http://starsteeds.blogspot.com and his Eclectarium blog can be accessed at http://eclectariumshuker.blogspot.com

There is also an entry for Dr Shuker in the online encyclopedia Wikipedia at http://en.wikipedia.org/wiki/Karl_Shuker and a Like (fan) page on Facebook.

AUTHOR BIBLIOGRAPHY

Mystery Cats of the World: From Blue Tigers To Exmoor Beasts (Robert Hale: London, 1989).

Extraordinary Animals Worldwide (Robert Hale: London, 1991).

The Lost Ark: New and Rediscovered Animals of the 20th Century (HarperCollins: London, 1993).

Dragons: A Natural History (Aurum: London/Simon & Schuster: New York, 1995; republished Taschen: Cologne, 2006).

In Search of Prehistoric Survivors: Do Giant 'Extinct' Creatures Still Exist? (Blandford: London, 1995).

The Unexplained: An Illustrated Guide to the World's Natural and Paranormal Mysteries (Carlton: London/JG Press: North Dighton, 1996; republished Carlton: London, 2002).

From Flying Toads To Snakes With Wings: From the Pages of FATE Magazine (Llewellyn: St Paul, 1997; republished Bounty: London, 2005).

Mysteries of Planet Earth: An Encyclopedia of the Inexplicable (Carlton: London, 1999).

The Hidden Powers of Animals: Uncovering the Secrets of Nature (Reader's Digest: Pleasantville/Marshall Editions: London, 2001).

The New Zoo: New and Rediscovered Animals of the Twentieth Century [fully-updated, greatly-expanded, second edition of *The Lost Ark*] (House of Stratus Ltd: Thirsk, UK/House of Stratus Inc: Poughkeepsie, USA, 2002).

The Beasts That Hide From Man: Seeking the World's Last Undiscovered Animals (Paraview: New York, 2003).

Extraordinary Animals Revisited: From Singing Dogs To Serpent Kings (CFZ Press: Bideford, 2007).

Dr Shuker's Casebook: In Pursuit of Marvels and Mysteries (CFZ Press: Bideford, 2008).

Dinosaurs and Other Prehistoric Animals on Stamps: A Worldwide Catalogue (CFZ Press: Bideford, 2008).

Star Steeds and Other Dreams: The Collected Poems (CFZ Press: Bideford, 2009).

Karl Shuker's Alien Zoo: From the Pages of Fortean Times (CFZ Press: Bideford, 2010).

The Encyclopaedia of New and Rediscovered Animals: From The Lost Ark to The New Zoo – and Beyond [fully-updated, greatly-expanded, third edition of *The Lost Ark*] (Coachwhip Publications: Landisville, 2012).

Cats of Magic, Mythology, and Mystery: A Feline Phantasmagoria (CFZ Press: Bideford, 2012).

Mirabilis: A Carnival of Cryptozoology and Unnatural History (Anomalist Books: New York, 2013).

Dragons in Zoology, Cryptozoology, and Culture (Coachwhip Publications: Greenville, 2013).

The Menagerie of Marvels: A Third Compendium of Extraordinary Animals (CFZ Press: Bideford, 2014).

Consultant and also Contributor

Man and Beast (Reader's Digest: Pleasantville, New York, 1993).
Secrets of the Natural World (Reader's Digest: Pleasantville, New York, 1993).
Almanac of the Uncanny (Reader's Digest: Surry Hills, Australia, 1995).
The Guinness Book of Records/Guinness World Records 1998-present day (Guinness: London, 1997-present day).

Consultant

Monsters (Lorenz: London, 2001).

Contributor

Of Monsters and Miracles CD-ROM (Croydon Museum/Interactive Designs: Oxton, 1995).
Fortean Times Weird Year 1996 (John Brown Publishing: London, 1996).
Mysteries of the Deep (Llewellyn: St Paul, 1998).
Guinness Amazing Future (Guinness: London, 1999).
The Earth (Channel 4 Books: London, 2000).
Mysteries and Monsters of the Sea (Gramercy: New York, 2001).
Chambers Dictionary of the Unexplained (Chambers: Edinburgh, 2007).
Chambers Myths and Mysteries (Chambers: Edinburgh, 2008).
The Fortean Times Paranormal Handbook (Dennis Publishing: London, 2009).
Plus numerous contributions to the annual *CFZ Yearbook* series of volumes.

Editor

Journal of Cryptozoology (the world's only peer-reviewed scientific journal devoted to cryptozoology, published by CFZ Press).

STILL ON THE TRACK OF UNKNOWN ANIMALS

The Centre for Fortean Zoology, or CFZ, is a non profit-making organisation founded in 1992 with the aim of being a clearing house for information, and coordinating research into mystery animals around the world.

We also study out of place animals, rare and aberrant animal behaviour, and Zooform Phenomena; little-understood "things" that appear to be animals, but which are in fact nothing of the sort, and not even alive (at least in the way we understand the term).

Not only are we the biggest organisation of our type in the world, but - or so we like to think - we are the best. We are certainly the only truly global cryptozoological research organisation, and we carry out our investigations using a strictly scientific set of guidelines. We are expanding all the time and looking to recruit new members to help us in our research into mysterious animals and strange creatures across the globe.

Why should you join us? Because, if you are genuinely interested in trying to solve the last great mysteries of Mother Nature, there is nobody better than us with whom to do it.

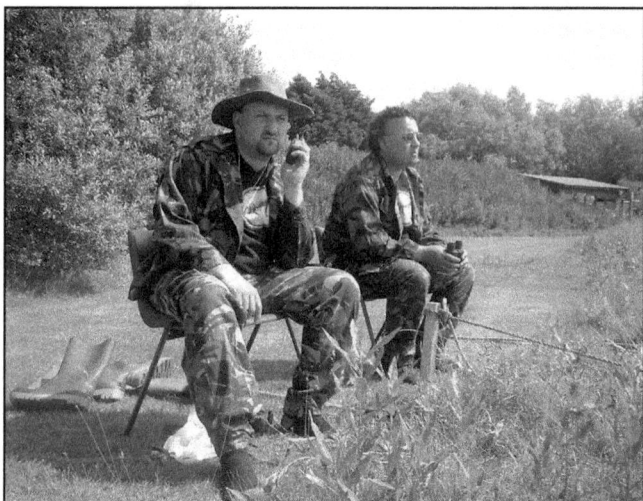

Members get a four-issue subscription to our journal *Animals & Men.* Each issue contains nearly 100 pages packed with news, articles, letters, research papers, field reports, and even a gossip column! The magazine is Royal Octavo in format with a full colour cover. You also have access to one of the world's largest collections of resource material dealing with cryptozoology and allied disciplines, and people from the CFZ membership regularly take part in fieldwork and expeditions around the world.

The CFZ is managed by a three-man board of trustees, with a non-profit making trust registered with HM Government Stamp Office. The board of trustees is supported by a Permanent Directorate of full and part-time staff, and advised by a Consultancy Board of specialists - many of whom are world-renowned experts in their particular field. We have regional representatives across the UK, the USA, and many other parts of the world, and are affiliated with

You'll find that the people at the CFZ are friendly and approachable. We have a thriving forum on the website which is the hub of an ever-growing electronic community. You will soon find your feet. Many members of the CFZ Permanent Directorate started off as ordinary members, and now work full-time chasing monsters around the world.

Write to us, e-mail us, or telephone us. The list of future projects on the website is not exhaustive. If you have a good idea for an investigation, please tell us. We may well be able to help.

We are always looking for volunteers to join us. If you see a project that interests you, do not hesitate to get in touch with us. Under certain circumstances we can help provide funding for your trip. If you look on the future projects section of the website, you can see some of the projects that we have pencilled in for the next few years.

In 2003 and 2004 we sent three-man expeditions to Sumatra looking for Orang-Pendek - a semi-legendary bipedal ape. The same three went to Mongolia in 2005. All three members started off merely subscribers to the CFZ magazine. Next time it could be you!

We have no magic sources of income. All our funds come from donations, membership fees, and sales of our publications and merchandise. We are always looking for corporate sponsorship, and other sources of revenue. If you have any ideas for fund-raising please let us know. However, unlike other cryptozoological organisations in the past, we do not live in an intellectual ivory tower. We are not afraid to get our hands dirty, and furthermore we are not one of those organisations where the membership have to raise money so that a privileged few can go on expensive foreign trips. Our research teams, both in the UK and abroad, consist of a mixture of experienced and inexperienced personnel. We are truly a community, and work on the premise that the benefits of CFZ membership are open to all.

Reports of our investigations are published on our website as soon as they are available. Preliminary reports are posted within days of the project finishing.

Each year we publish a 200 page yearbook

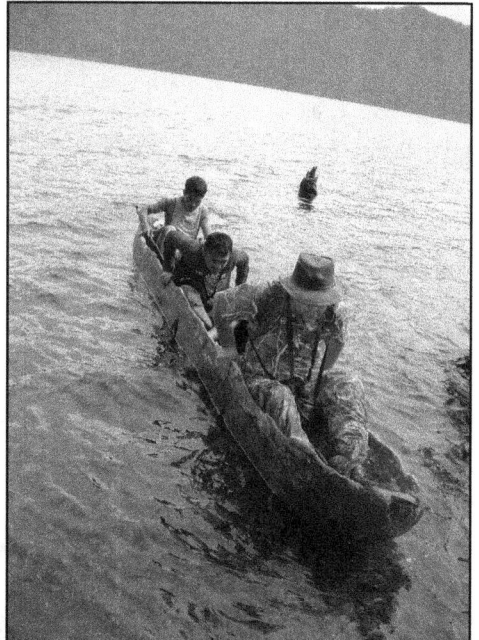

We have a thriving YouTube channel, CFZtv, which has well over two hundred self-made documentaries, lecture appearances, and episodes of our monthly webTV show. We have a daily online magazine, which has over a million hits each year.

Each year since 2000 we have held our annual convention - the Weird Weekend. It is three days of lectures, workshops, and excursions. But most importantly it is a chance for members of the CFZ to meet each other, and to talk with the members of the permanent directorate in a relaxed and informal setting and preferably with a pint of beer in one hand. Since 2006 - the Weird Weekend has been bigger and better and held on the third weekend in August in the idyllic rural location of Woolsery in North Devon.

Since relocating to North Devon in 2005 we have become ever more closely involved with other community organisations, and we hope that this trend will continue. We have also worked closely with Police Forces across the UK as consultants for animal mutilation cases, and we intend to forge closer links with the coastguard and other community services. We want to work closely with those who regularly travel into the Bristol Channel, so that if the recent trend of exotic animal visitors to our coastal waters continues, we can be out there as soon as possible.

Apart from having been the only Fortean Zoological organisation in the world to have consistently published material on all aspects of the subject for over a decade, we have achieved the following concrete results:

• Disproved the myth relating to the headless so-called sea-serpent carcass of Durgan beach in Cornwall 1975
• Disproved the story

of the 1988 puma skull of Lustleigh Cleave

- Carried out the only in-depth research ever into the mythos of the Cornish Owlman.
- Made the first records of a tropical species of lamprey
- Made the first records of a luminous cave gnat larva in Thailand
- Discovered a possible new species of British mammal - the beech marten
- In 1994-6 carried out the first archival fortean zoological survey of Hong Kong
- In the year 2000, CFZ theories were confirmed when a new species of lizard was added to the British List
- Identified the monster of Martin Mere in Lancashire as a giant wels catfish
- Expanded the known range of Armitage's skink in the Gambia by 80%
- Obtained photographic evidence of the remains of Europe's largest known pike
- Carried out the first ever in-depth study of the ninki-nanka
- Carried out the first attempt to breed Puerto Rican cave snails in captivity
- Were the first European explorers to visit the `lost valley` in Sumatra
- Published the first ever evidence for a new tribe of pygmies in Guyana
- Published the first evidence for a new species of caiman in Guyana

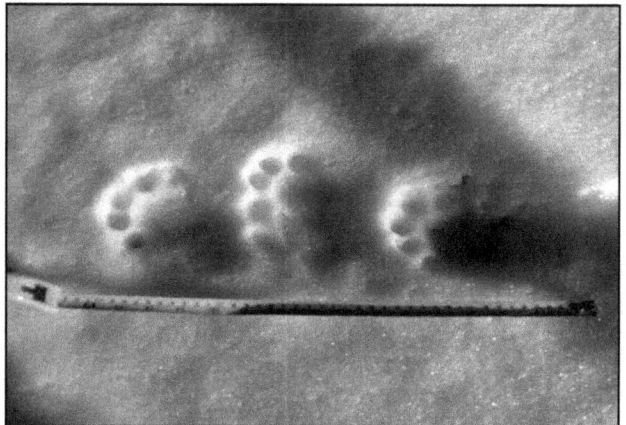

We have a thriving YouTube channel, CFZtv, which has well over two hundred self-made documentaries, lecture appearances, and episodes of our monthly webTV show. We have a daily online magazine, which has over a million hits each year.

Each year since 2000 we have held our annual convention - the Weird Weekend. It is three days of lectures, workshops, and excursions. But most importantly it is a chance for members of the CFZ to meet each other, and to talk with the members of the permanent directorate in a relaxed and informal setting and preferably with a pint of beer in one hand. Since 2006 - the Weird Weekend has been bigger and better and held on the third weekend in August in the idyllic rural location of Woolsery in North Devon.

Since relocating to North Devon in 2005 we have become ever more closely involved with other community organisations, and we hope that this trend will continue. We have also worked closely with Police Forces across the UK as consultants for animal mutilation cases, and we intend to forge closer links with the coastguard and other community services. We want to work closely with those who regularly travel into the Bristol Channel, so that if the recent trend of exotic animal visitors to our coastal waters continues, we can be out there as soon as possible.

Apart from having been the only Fortean Zoological organisation in the world to have consistently published material on all aspects of the subject for over a decade, we have achieved the following concrete results:

- Disproved the myth relating to the headless so-called sea-serpent carcass of Durgan beach in Cornwall 1975
- Disproved the story

of the 1988 puma skull of Lustleigh Cleave

- Carried out the only in-depth research ever into the mythos of the Cornish Owlman.
- Made the first records of a tropical species of lamprey
- Made the first records of a luminous cave gnat larva in Thailand
- Discovered a possible new species of British mammal - the beech marten
- In 1994-6 carried out the first archival fortean zoological survey of Hong Kong
- In the year 2000, CFZ theories were confirmed when a new species of lizard was added to the British List
- Identified the monster of Martin Mere in Lancashire as a giant wels catfish
- Expanded the known range of Armitage's skink in the Gambia by 80%
- Obtained photographic evidence of the remains of Europe's largest known pike
- Carried out the first ever in-depth study of the ninki-nanka
- Carried out the first attempt to breed Puerto Rican cave snails in captivity
- Were the first European explorers to visit the `lost valley` in Sumatra
- Published the first ever evidence for a new tribe of pygmies in Guyana
- Published the first evidence for a new species of caiman in Guyana

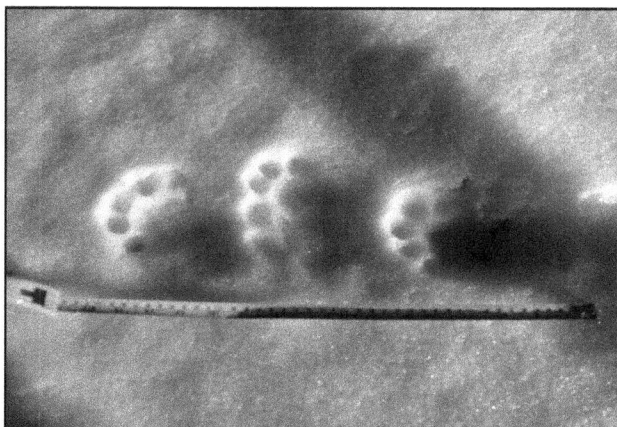

- 2006 Gambia (Gambo - Gambian sea monster , Ninki Nanka and Armitage's skink
- 2006 Llangorse Lake (Giant pike, giant eels)
- 2006 Windermere (Giant eels)
- 2007 Coniston Water (Giant eels)
- 2007 Guyana (Giant anaconda, didi, water tiger)
- 2008 Russia (Almasty)
- 2009 Sumatra (Orang pendek)
- 2009 Republic of Ireland (Lake Monster)
- 2010 Texas (Blue Dogs)
- 2010 India (Mande Burung)
- 2011 Sumatra (Orang-pendek)
- 2013 Sumatra (Orang Pendek)
- 2013 Tasmania (Thylacine)

For details of current membership fees, current expeditions and investigations, and voluntary posts within the CFZ that need your help, please do not hesitate to contact us.

The Centre for Fortean Zoology,
Myrtle Cottage,
Woolfardisworthy,
Bideford, North Devon
EX39 5QR

Telephone 01237 431413
Fax+44 (0)7006-074-925
eMail info@cfz.org.uk

Websites:

www.cfz.org.uk
www.weirdweekend.org

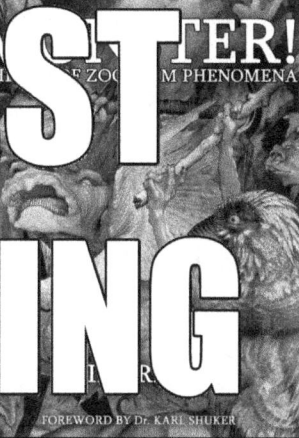

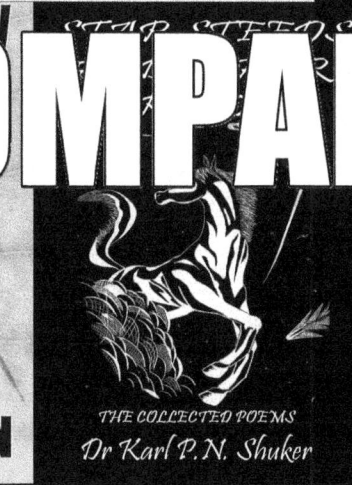

THE WORLD'S WEIRDEST PUBLISHING COMPANY

HOW TO START A PUBLISHING EMPIRE

Unlike most mainstream publishers, we have a non-commercial remit, and our mission statement claims that "we publish books because they deserve to be published, not because we think that we can make money out of them". Our motto is the Latin Tag *Pro bona causa facimus* (we do it for good reason), a slogan taken from a children's book *The Case of the Silver Egg* by the late Desmond Skirrow.

WIKIPEDIA: "The first book published was in 1988. *Take this Brother may it Serve you Well* was a guide to Beatles bootlegs by Jonathan Downes. It sold quite well, but was hampered by very poor production values, being photocopied, and held together by a plastic clip binder. In 1988 A5 clip binders were hard to get hold of, so the publishers took A4 binders and cut them in half with a hacksaw. It now reaches surprisingly high prices second hand.

The production quality improved slightly over the years, and after 1999 all the books produced were ringbound with laminated colour covers. In 2004, however, they signed an agreement with Lightning Source, and all books are now produced perfect bound, with full colour covers."

Until 2010 all our books, the majority of which are/were on the subject of mystery animals and allied disciplines, were published by `CFZ Press`, the publishing arm of the Centre for Fortean Zoology (CFZ), and we urged our readers and followers to draw a discreet veil over the books that we published that were completely off topic to the CFZ.

However, in 2010 we decided that enough was enough and launched a second imprint, `Fortean Words` which aims to cover a wide range of non animal-related esoteric subjects. Other imprints will be launched as and when we feel like it, however the basic ethos of the company remains the same: Our job is to publish books and magazines that we feel are worth publishing, whether or not they are going to sell. Money is, after all - as my dear old Mama once told me - a rather vulgar subject, and she would be rolling in her grave if she thought that her eldest son was somehow in `trade`.

Luckily, so far our tastes have turned out not to be that rarified after all, and we have sold far more books than anyone ever thought that we would, so there is a moral in there somewhere…

Jon Downes,
Woolsery, North Devon
July 2010

CFZ PRESS

Wildman! by Redfern, Nick
Globsters by Newton, Michael
Cats of Magic, Mythology and Mystery Shuker, by Karl P. N
Those Amazing Newfoundland Dogs by Bondeson, Jan
The Mystery Animals of Pennsylvania by Gable, Andrew
Sea Serpent Carcasses - Scotland from the Stronsa Monster to Loch Ness by Glen Vaudrey
The CFZ Yearbook 2012 edited by Jonathan and Corinna Downes
ORANG PENDEK: Sumatra's Forgotten Ape by Richard Freeman
THE MYSTERY ANIMALS OF THE BRITISH ISLES: London by Neil Arnold
CFZ EXPEDITION REPORT: India 2010 by Richard Freeman *et al*
The Cryptid Creatures of Florida by Scott Marlow
Dead of Night by Lee Walker
The Mystery Animals of the British Isles: The Northern Isles by Glen Vaudrey
THE MYSTERY ANIMALS OF THE BRTISH ISLES: Gloucestershire and Worcestershire by
Paul Williams
When Bigfoot Attacks by Michael Newton
Weird Waters – The Mystery Animals of Scandinavia: Lake and Sea Monsters by Lars Thomas
The Inhumanoids by Barton Nunnelly
Monstrum! A Wizard's Tale by Tony "Doc" Shiels
CFZ Yearbook 2011 edited by Jonathan Downes
Karl Shuker's Alien Zoo by Shuker, Dr Karl P.N
Tetrapod Zoology Book One by Naish, Dr Darren
The Mystery Animals of Ireland by Gary Cunningham and Ronan Coghlan
Monsters of Texas by Gerhard, Ken
The Great Yokai Encyclopaedia by Freeman, Richard
NEW HORIZONS: Animals & Men issues 16-20 Collected Editions Vol. 4
by Downes, Jonathan
A Daintree Diary -
Tales from Travels to the Daintree Rainforest in tropical north Queensland, Australia
by Portman, Carl
Strangely Strange but Oddly Normal by Roberts, Andy

by Downes, Jonathan
The Smaller Mystery Carnivores of the Westcountry by Downes, Jonathan
CFZ EXPEDITION REPORT: Gambia 2006 by Richard Freeman *et al*, Shuker, Karl (fwd)
The Owlman and Others by Jonathan Downes
The Blackdown Mystery by Downes, Jonathan
Big Cats in Britain Yearbook 2006 by Fraser, Mark (Ed)
Fragrant Harbours - Distant Rivers by Downes, John T
Only Fools and Goatsuckers by Downes, Jonathan
Monster of the Mere by Jonathan Downes
Dragons:More than a Myth by Freeman, Richard Alan
Granfer's Bible Stories by Downes, John Tweddell
Monster Hunter by Downes, Jonathan

CFZ Classics is a new venture for us. There are many seminal works that are either unavailable today, or not available with the production values which we would like to see. So, following the old adage that if you want to get something done do it yourself, this is exactly what we have done.

Desiderius Erasmus Roterodamus (b. October 18th 1466, d. July 2nd 1536) said: "When I have a little money, I buy books; and if I have any left, I buy food and clothes," and we are much the same. Only, we are in the lucky position of being able to share our books with the wider world. CFZ Classics is a conduit through which we cannot just re-issue titles which we feel still have much to offer the cryptozoological and Fortean research communities of the 21st Century, but we are adding footnotes, supplementary essays, and other material where we deem it appropriate.

Headhunters of The Amazon by Fritz W Up de Graff (1902)

Fortean Words

The Centre for Fortean Zoology has for several years led the field in Fortean publishing. CFZ Press is the only publishing company specialising in books on monsters and mystery animals. CFZ Press has published more books on this subject than any other company in history and has attracted such well known authors as Andy Roberts, Nick Redfern, Michael Newton, Dr Karl Shuker, Neil Arnold, Dr Darren Naish, Jon Downes, Ken Gerhard and Richard Freeman.

Now CFZ Press are launching a new imprint. Fortean Words is a new line of books dealing with Fortean subjects other than cryptozoology, which is - after all - the subject the CFZ are best known for.

Other books include a look at the Berwyn Mountains UFO case by renowned Fortean Andy Roberts and a series of forthcoming books by transatlantic researcher Nick Redfern. CFZ Press are dedicated to maintaining the fine quality of their works with Fortean Words. New authors tackling new subjects will always be encouraged, and we hope that our books will continue to be as ground-breaking and popular as ever.

Haunted Skies Volume One 1940-1959 by John Hanson and Dawn Holloway
Haunted Skies Volume Two 1960-1965 by John Hanson and Dawn Holloway
Haunted Skies Volume Three 1965-1967 by John Hanson and Dawn Holloway
Haunted Skies Volume Four 1968-1971 by John Hanson and Dawn Holloway
Haunted Skies Volume Five 1972-1974 by John Hanson and Dawn Holloway
Haunted Skies Volume Six 1975-1977 by John Hanson and Dawn Holloway
Grave Concerns by Kai Roberts

Police and the Paranormal by Andy Owens
Dead of Night by Lee Walker
Space Girl Dead on Spaghetti Junction - an anthology by Nick Redfern
I Fort the Lore - an anthology by Paul Screeton
UFO Down - the Berwyn Mountains UFO Crash by Andy Roberts
The Grail by Ronan Coghlan
UFO Warminster - Cradle of Contract by Kevin Goodman
Quest for the Hexham Heads by Paul Screeton

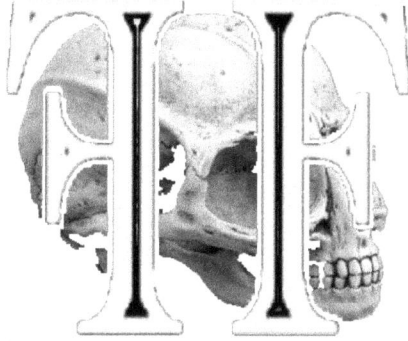

Fortean Fiction

Just before Christmas 2011, we launched our third imprint, this time dedicated to - let's see if you guessed it from the title - fictional books with a Fortean or cryptozoological theme. We have published a few fictional books in the past, but now think that because of our rising reputation as publishers of quality Forteana, that a dedicated fiction imprint was the order of the day.

We launched with four titles:

Green Unpleasant Land by Richard Freeman
Left Behind by Harriet Wadham
Dark Ness by Tabitca Cope
Snap! By Steven Bredice
Death on Dartmoor by Di Francis
Dark Wear by Tabitca Cope
Hyakymonogatari Book 1 by Richard Freeman

www.ingramcontent.com/pod-product-compliance
Lightning Source LLC
Chambersburg PA
CBHW060838280326
41934CB00007B/829